SpringerBriefs on Pioneers in Science and Practice

Texts and Protocols

Volume 13

Series Editor

Hans Günter Brauch

For further volumes:
http://www.springer.com/series/11446
http://www.afes-press-books.de/html/SpringerBriefs_PSP_TP.htm

Arthur H. Westing

From Environmental to Comprehensive Security

Springer

Arthur H. Westing
Westing Associates in Environment, Security,
& Education
Putney, VT
USA

ISSN 2194-3125 ISSN 2194-3133 (electronic)
ISBN 978-3-319-00686-4 ISBN 978-3-319-00687-1 (eBook)
DOI 10.1007/978-3-319-00687-1
Springer Cham Heidelberg New York Dordrecht London

Library of Congress Control Number: 2013939044

Printed on acid-free paper

Springer is part of Springer Science+Business Media (www.springer.com)

Preface

It was a special honor for me to start this new book series on *Pioneers in Science and Practice* (PSP) with a volume on *Arthur H. Westing*: *Pioneer on the Environmental Impact of War* and to be able to introduce him as such in that book (Westing 2013). Now I am most pleased to have been able to convince Arthur to prepare this second volume, to coincide with his 85th birthday in July 2013, a volume that brings together many of his innovative contributions since the early 1980s on the 'reconceptualization of security', with a particular focus on environmental and comprehensive security.

Thus, Arthur has not only been a pioneer on the environmental impact of war, but also a major innovative contributor since the 1980s to a conceptual scientific discourse and policy debate on a 'reconceptualization of security' at a time when the Cold War was winding down and that has been underway ever since (cf. Brauch et al. 2008, 2009, 2011). Elsewhere I have distinguished among three processes in the broadening of the dimensions of security from the narrow politicomilitary: (a) to a *widening* of its societal, economic, and environmental or ecological characteristics; (b) to a *deepening* from a state-centered to a people-centered security, that is, from the nation-state as the referent object to one in which human beings, communities, and humankind are the referent objects—a concept to be framed as 'human security'; and (c) to a *sectorialization* of security to encompass the notions of energy, food, water, soil, and health security.

Arthur's innovative contribution was primarily to the 'widening' of security by bringing—as a forest ecologist and environmental scientist—environmental considerations into the evolving policy debate on national and international security. He addressed simultaneously the two sides of the same coin: (a) of the environmental impact of war (Westing 2013); and (b) of the environmental causes of or impacts on multiple forms of conflict (cf. Chap. 1). He framed 'environmental security' initially as a state-centered concept and gradually moved to what he called 'comprehensive human security' (Westing 2013, pp 15–17), taking up recent critical suggestions (cf. Bogardi and Brauch 2005; Brauch 2005a, b, 2008, 2011; Brauch and Scheffran 2012; Dalby et al. 2009).

In Chap. 2 Arthur offers us a chronological bibliographical reference list to his many contributions to the reconceptualization of security from 1981 until 2013. Indeed, several of his early contributions inspired me to launch a global multidisciplinary project on the 'Reconceptualization of Security'. He introduced environmental factors to national, international, and regional security (cf. Chaps. 4–7). Arthur used the term 'human security' in a United Nations Environment Programme publication of 1993, a year before Mabhuqul Haq triggered a global debate on human security in the United Nations Development Programme's annual report on human development. As a forest scientist, Arthur was familiar with the debates in the environmental, security, and peace research communities and could thus contribute to conceptual bridge building among those diverse fields.

Arthur's conceptualization of environmental security in Chap. 4 is still very relevant. For Arthur 'regional security' has always been 'an ecological necessity' (cf. Chap. 4). He applied it to the analysis of 'maritime issues' (cf. Chap. 5) and to 'transfrontier cooperation' (cf. Chap. 6), also including a debate on biodiversity issues and the role of protected areas that have been under-researched in the environmental security debate.

I met Arthur most recently in November 2010 at an international conference in Berlin on 'A World Without Walls' where he presented a paper on the 'Korean Demilitarized Zone (DMZ) as a Bridge Between the Two Koreas' (cf. Chap. 7) where he began by comparing several divided countries (Germany, Yemen, Korea). He suggested then to protect the DMZ as a 'confidence and security building measure', a proposal he developed into a proposed legal text for a treaty between both Koreas.

In Chap. 8 Arthur addresses linkages between globalization and environmental security; and in Chap. 9 he has examined in-depth the concept and significance of 'environmental refugees', which emerged from the first conference on Desertification and Migration held in Almeria, Spain in 1994. I read this text when I prepared my opening speech to the second conference on Desertification and Migration held in Almeria in 2006 that firmly placed this linkage of desertification and refugees on the environmental security agendas of nation-states, the United Nations Environment Programme, the United Nations Development Programme, and the United Nations Convention to Combat Desertification.

In Chap. 10 Arthur discusses population as perhaps *the* basic issue, using the concept of the 'global carrying capacity', having been inspired, among others, by Lester Brown, Garrett Hardin, and Paul Ehrlich, thus relying more on the ecologist discourse and less on the debates among demographers who stress fertility, mortality, and the like while leaving out environmental constraints and environmentally induced migration. Thus, Arthur has once again contributed to a conceptual bridge building that is essential for understanding the global environmental challenges of the twenty-first century.

The present volume (PSP-13) reintroduces the innovative work of a major scientific, conceptual bridge-builder who has applied multidisciplinary and interdisciplinary approaches in linking his experience as a forest scientist with debates on the environment, war, and demography; and at the same time often addressing the legal dimension. This specific perspective has contributed what biologist Edward O. Wilson (1998) has called *consilience* (the interlocking of causal explanations across disciplines) in which the 'interfaces between disciplines become as important as the disciplines themselves' that would 'touch the borders of the social sciences and humanities'.

On the occasion of Arthur H. Westing's 85th birthday in July 2013, both volumes (PSP-1, PSP-13) make a part of Arthur's contribution to Wilson's innovative concept of 'consilience' widely accessible to students, whether in hard copy or as electronic publications (via laptop, tablet, computer, or smartphone). These two volumes can be considered as 'intellectual appetizers' that will in turn guide the reader to Arthur's major books, chapters, and articles. I would suggest these two texts to be ideal for course adoption in graduate seminars on environmental, security, peace, and development issues around the world. Indeed, some 4,500 universities and academic institutions globally subscribe to Springer's earth science package, enabling students to download both of these books at no cost.

Arthur's many friends and colleagues around the globe wish him good health and continued creativity. I have learnt much from Arthur's conceptual work and was influenced by his contributions across disciplinary boundaries. As a series editor, my intention has been to share Arthur's experience with future generations and encourage them also to overcome the methodological, theoretical, and disciplinary confinements that prevail in highly specialized journals and that make it more difficult to communicate the scientific results to policymakers and international organizations. Arthur's work at the Stockholm International Peace Research Institute and the International Peace Research Institute Oslo during the 1970s and 1980s were sponsored and fully supported by the United Nations Environment Programme's Executive Director Mustafa Tolba (who, it might be mentioned, withstood US pressure to stop funding of Arthur's UNEP work).

In short, Arthur has been a *conceptual and empirical innovator* in his several roles during the six decades of his professional life: (a) as a *natural scientist*; (b) as a *professor and educator*; (c) as a *concerned US citizen*; (d) as an *ecologist and environmentalist*; (e) as a *peace researcher*; (f) as a *policy consultant*; and (g) as a *politically active scientist and citizen*. He thus became a *Vorbild* for many young scientists, but also for policymakers, to take the courage not to remain silent on the misuse of scientific knowledge in warfare or for increasing short-term economic benefits and ignoring the longer term effects on the life of present and future generations. This is what the notion of 'sustainable development' is all about, as suggested by the Brundtland Commission (Brundtland et al. 1987). Arthur needs many followers to move toward policies of a transition to sustainability during this century to avoid the security impacts of global environmental change, of biodiversity loss, and of chaotic interactions within the earth system that could result in tipping points endangering the lives and livelihoods of millions of people. Arthur's two books put the analysis of scientific linkages on the agenda of course planners and of policymakers to facilitate moving from 'knowledge to action'.

Cuernavaca, Mexico, April 2013

Hans Günter Brauch
Editor, Springer Briefs on Pioneers
in Science and Practice

Hans Günter Brauch (Dr. phil. habil.) has taught as a Privatdozent (Adjunct Professor) at the Free University of Berlin from 1999 until his retirement in 2012. Since 1987 he has been Chairman of Peace Research and European Security Studies (AFES-PRESS). Since 2003 he has been Editor of the peer-reviewed *Hexagon Series on Human and Environmental Security and Peace* (HESP), and since 2012 he is the Editor of the *Springer Briefs in Environment, Security, Development and Peace* (ESDP) as well as of this *Series on Pioneers in Science and Practice* (PSP). In 2012 he taught at SciencePo (Paris), in the Ph.D. Programme of the *Centro de Estudios Superiores Navales* (CESNAV) in Mexico, at Arhus University (Denmark), at the *European Peace University* (EPU) (Austria), at the *National University of Malaysia* (UKM), and at Chulalongkorn University (Bangkok).

References

Bogardi, J. & Brauch, H. G. (2005). Global environmental change: A challenge for human security—defining and conceptualising the environmental dimension of human security. In: A. Rechkemmer, (ed.) *UNEO: Towards an International Environment Organization: Approaches to a Sustainable Reform of Global Environmental Governance*. Baden–Baden, Germany: Nomos, 310 pp: (pp. 85–109).

Brauch, H. G. (2005a). *Environment and Human Security: Towards Freedom from Hazard Impacts*. Bonn, Germany: UN University, Institute for Environment and Human Security, InterSecTions No. 2/2005, pp. 56 http://www.ehs.unu.edu/file/get/4031.

Brauch, H. G. (2005b). *Threats, Challenges, Vulnerabilities and Risks in Environmental and Human Security*. Bonn, Germany: UN University, Institute for Environment and Human Security, Source No. 1/2005, pp. 100 http://www.ehs.unu.edu/file/get/4040.

Brauch, H. G. (2008). Conceptualising the environmental dimension of human security in the UN. *International Social Science Journal Monograph Series* (Paris), 148 pp: pp 19–48. [This Monograph was edited by M. Goucha & J. Crowley under the title *Rethinking Human Security* and published in September 2008].

Brauch, H. G. (2011). *The Environmental Dimension of Human Security—Freedom from Hazard Impacts*. New York: UN General Assembly, Interactive Debate 2: Human Security—Its Application and Added Value, 6 pp. 14 April 2011. http://www.un.org/en/ga/president/65/initiatives/Human%20Security/DrBrauch.pdf & http://www.un.org/News/Press/docs//2011/ga11072.doc.htm.

Brauch, H. G. et al. (eds). (2008). *Globalization and Environmental Challenges: Reconceptualizing Security in the 21st Century*. pp. 1147 Berlin: Springer.

Brauch, H. et al. (eds). (2009). *Facing Global Environmental Change: Environmental, Human, Energy, Food, Health and Water Security Concepts* pp. 1586. Berlin: Springer.

Brauch, H.G. et al. (eds). (2011). *Coping with Global Environmental Change, Disasters and Security: Threats, Challenges, Vulnerabilities and Risks* pp. 1815. Berlin: Springer.

Brauch, H. G. & Scheffran, J. (2012). Introduction: climate change, human security, and violent conflict in the Anthropocene. In: Scheffran, J. et al. (eds): *Climate Change, Human Security and Violent Conflict: Challenges for Societal Stability* 868 pp. 1–40. Heidelberg: Springer.

Brundtland, G. H. et al. (1987). *Our Common Future: World Commission on Environment and Development*. rev edn. Oxford: Oxford University Press, pp. 400.

Dalby, S. et al. (2009). Environmental security concepts revisited during the first three phases (1983–2006). In: H. G. Brauch et al. (eds). *Facing Global Environmental Change: Environmental, Human, Energy, Food, Health and Water Security Concepts*. 1586 pp. 781–790, Berlin: Springer.

Westing, A. H. (2013). *Arthur H. Westing*: *Pioneer on the Environmental Impact of War*. Heidelberg: Springer, 150 pp. (Springer Briefs on Pioneers in Science and Practice, Vol. 1) http://www.afes-press-books.de/html/SpringerBriefs_PSP01.htm & http://www.springer.com/law/environmental/book/978-3-642-31321-9.

Wilson, E. O. (1998). *Consilience: the Unity of Knowledge* pp. 332 New York: Knopf.

Arthur H. Westing: A Personal Memoir

Arthur Westing came to the International Peace Research Institute Oslo (PRIO) in January 1988. Sverre Lodgaard, who had worked at the Stockholm International Peace Research Institute (SIPRI) for the past 6 years, returned to Oslo to take over the position as Director of PRIO. He was able to bring Arthur to Oslo at the same time, along with his project on 'Peace, Security, and Environment' funded by the United Nations Environment Programme (UNEP). SIPRI's loss was PRIO's gain. At PRIO Arthur edited a volume on *Environmental Hazards of War*, which dealt with the planned or inadvertent release of pollutants following the destruction of major industries in war. Before that, however, he had edited another volume, *Comprehensive Security for the Baltic*, which focused on security in the Baltic region as seen through the lens of an extended concept of security. Arthur did not invent the concept of environmental security. But his work was (and remains) one of the most thorough and thoughtful expositions of it.

When Arthur went back to the USA (to retire, I thought—some retirement!) I had to take over the organizing of a UNEP-funded conference in 1991 on conversion and the environment which had fallen between the cracks. Some of the UNEP-funded conferences had to be held in the USSR in order to spend some of the non-convertible rubles in which that country had paid a major portion of its dues. I have to confess that I took on this project largely because I was fascinated by the opportunity to visit Perm (known as Molotov in my school days), a city closed to foreigners until just before our conference was held there. But my academic interest in the relationship between the environment and security arose from that experience, and continues to this day.

Eventually, my colleagues and I came to focus more on the environmental causes of armed conflict than on its consequences. We have probably taken the work in a direction more critical of neomalthusian thinking than Arthur might have felt comfortable with. I am somewhat less pessimistic than Arthur on several issues discussed in this volume, such as the risk of transboundary atmospheric pollution leading to international conflict (cf. page 20), that the quest for human security may have become more elusive (cf. page 63), that economic globalization may be harmful to environmental security (cf. page 114), that environmental security is seriously deteriorating at the global level, and that global overpopulation is our most serious problem (cf. page 133). But if you want an intelligent defense of those propositions, you can do a lot worse than to consult this volume.

One of my most profound memories of Arthur's work at PRIO relates to his extreme attention to detail and accuracy, which the reader will soon discover in this volume too. Arthur found errors even in reputable collections of treaties and other standard works of reference. One of the few people who seemed able to live up to Arthur's high standards of citations and references was his valued erstwhile colleague from SIPRI, the recently deceased Jozef Goldblat. With an office right next to Arthur's I couldn't help noticing the occasional outburst when someone did not meet his exacting standards. In one of my papers I inadvertently cited him as 'Arthur F. Westing'. This was not easily forgiven. Nor should it be.

But I thought even Arthur had gone too far when I learned that he was asking every author who had a direct quotation in a chapter in one of his books to send him a photocopy of that quote from the original publication. Surely this was going too far! Shortly thereafter I was in charge of an edited volume myself. PRIO's discerning copy editor pointed out some language infelicities that appeared to have been committed by prominent writers—if one were to believe the lesser mortals who had cited them. Did they really say that? I was too embarrassed to ask the authors to send me copies of the originals, so I looked them up myself. And, indeed, there were numerous errors. Of course, scholars often copy quotations and references from previous articles they have read, so any errors get reproduced. Most of them are trivial, but once in a while there will be one 'not' too many (or one too few). Those of my colleagues at PRIO who think I've spent too much time correcting details now know where I learned that *modus operandi*.

Many of my fruitful interactions with Arthur were linked to my role as Editor of the *Journal of Peace Research*. Arthur published two articles and several short book reviews in the *Journal* and I took advantage of his presence to solicit referee reports from him on many occasions. In fact, over a 15-year period, he was among the top 5 % of referees in terms of the number of reports. I was particularly impressed by the fact that, as a matter of principle, Arthur always signed his reviews with his full name. Many scholars, however committed they may be to transparency, are reluctant to do this because it exposes them to potential quarrelsome responses from authors who felt that their fine scholarship had not been sufficiently appreciated by the *Journal*.

On factual academic matters, I was never able to catch Arthur out. My moment of triumph arrived years later when my son and I stayed with Arthur and Carol over a weekend in Vermont and went hiking with them in a local nature reserve that they had helped to establish, reflecting their environmental activism. This was in the fall and I suggested bringing along a basket for picking mushrooms. There are no edible mushrooms in that area, Arthur stated with some finality. Since my stubbornness matches his, I brought the basket along anyway. That evening we all had mushrooms at supper, and no one got sick. I was later told by Carol that no sooner had I left than Arthur went off to buy a mushroom field guide. This was a dozen or so years ago. He hadn't stopped taking in new knowledge then. He hasn't stopped at 85. And he never will.

8 April 2013

Nils Petter Gleditsch
Research Professor at PRIO and
Professorof Political Science at the
Norwegian University of Science and Technology

Contents

Part I
Toward a Sustainable Future

Chapter 1
National and International Security: An Evolving Concept

1.1 Beyond the Environmental Impact of War

The present volume serves as an extension of my studies that have dealt with the environmental impact of war, the subject of the first volume in this series of 'Springer Briefs on Pioneers in Science and Practice' (Westing 2013). As already suggested there, I have had a lifelong affinity with the outside world and the plants and animals it contains (Westing 2013: 4).[1]

My own awareness of the environmental disruption that is so often inflicted upon the battlefield (and also more subtly further afield, and even during times of peace) was made amply clear by my own contributions to such damage during the Korean War of 1950–1953. And that in turn had been subsequently reinforced especially by my on-site studies during the Second Indochina War of 1961–1975, a

This Chapter was prepared *de novo*. The **Appendix** to this Chapter was prepared by the author while serving as a Consultant to the Eritrean Agency for the Environment, and adopted by the Government of Eritrea in February 1995; it is reproduced from the author's Entry #282 provided in Chap. 2, with the original title, 'Eritrean National Code of Conduct for Environmental Security'. It is not copyrighted by the Government of Eritrea; it is used here by permission of Westing Associates in Environment, Security, & Education, the copyright holder, as given on 13 March 2013. The author is pleased to acknowledge highly useful discussions with Naigzy Gebremedhin.

[1] My formal training was in botany (B.A., 1950, Columbia University), in silviculture (M.F., 1954, Yale University), and in plant autecology (Ph.D., 1959, Yale University). I have been a Research Forester with the US Forest Service, working on silvicides; and variously a Professor of Forestry, of Ecology, and of Conservation (at Purdue University, the University of Massachusetts, Middlebury College, Windham College, Hampshire College, and the European Peace University). I have been a Senior Researcher at the Stockholm International Peace Research Institute and the International Peace Research Institute Oslo; and also a Consultant to the United Nations Environment Programme as well as other United Nations agencies. As to my military background, I have been an infantry and artillery officer (Second Lieutenant to Captain) in the US Marine Corps that included about a year of continuous combat duty during the Korean War of 1950–1953.

A. H. Westing, *From Environmental to Comprehensive Security*,
SpringerBriefs on Pioneers in Science and Practice 13,
DOI: 10.1007/978-3-319-00687-1_1, © The Author(s) 2013

war in which deliberate environmental and related agricultural devastation was a major component of the US strategy to subdue its elusive enemy (Westing 2013: 35–49)—with those Vietnamese studies having gained scientific validity owing in part to my prior US Forest Service research activities on the chemical killing of unwanted trees.

Then, as I became ever more aware of, and concerned over, the overall damage being inflicted upon the global biosphere by the civil sectors in both the highly developed Industrialized World and the less developed Third World, I began to hope that the situation could be at least somewhat alleviated if only the military damage could be substantially reduced. Although warfare has been an unfailingly continuous human activity throughout the entire existence of our species, it has nonetheless been my hope that with the continually worsening global environment, and as sapient creatures, we could finally learn to resolve at least most of our many conflicts without resort to arms. And if not, to at least learn to wage war with less damage to the environment.

Turning now more specifically to **security**, the subject of this volume, '**national security**' has traditionally been defined as referring to the ability of a country to defend itself militarily from the threat or actuality of an attack coming from either without or within. But then a seminal publication in 1977 by Lester Brown of the Worldwatch Institute (Washington) introduced (to me at least) the notion of '**environmental security**' as being a most necessary additional component of national security (Brown 1977). In following up on this conceptual expansion in my own thinking, I quickly recognized that by rights it was the solemn obligation for a country not only to provide national military plus environmental security for its citizenry, but also '**social (societal) security**'. And even more, important, I recognized that in order to provide such national social security it was necessary to simultaneously provide national environmental security, with neither of those two securities achievable unless *both* were. I published my early thoughts on this matter in April 1983 (Westing 1983), and then soon began expanding on and fleshing out my understanding of the idea (e.g., Westing 1986; Westing 1989; cf. also Chap. 2). In other words, it is important to reiterate that the actions necessary for our leaders and fellow citizens to pursue were little different no matter whether one's primary concern had an anthropocentric basis or an ecocentric one.

I was, of course, not alone in the 1980s in further exploring the notion of environmental security. Influential scholars early on included Richard Ullman of Princeton University (Ullman 1983–1984); Nansen Behar of the Bulgarian Academy of Sciences (Behar 1985); Naigzy Gebremedhin of UNEP together with Sverre Lodgaard of PRIO, Johannes Opschoor of the Free University of Amsterdam, and Renat Perelet of the Institute of Systems Studies in Moscow (Gebremedhin et al. 1989); and Jessica Matthews of the World Resources Institute in Washington (Matthews 1989). Moreover, the related concept of 'sustainable development' was put forth by the International Union for Conservation of Nature in 1980 (IUCN 1980) and further developed by the *ad hoc* World Commission on Environment and Development (Brundtland et al. 1987).[2]

Inasmuch as political boundaries seldom coincide with ecological boundaries, it did not take me long to grasp that for a country to achieve environmental security it often had to act in cooperation with neighboring countries. This began my interest in **transfrontier** (cross-border) cooperation—something I was hoping would not only benefit the environment of what I termed the 'ecogeographical region' (whether binational or multinational), but might at the same time serve as a political confidence-building measure between the involved neighbors.

I soon also began to realize that most countries, and our globe as a whole, simply had too many people to make it possible to meet the basic needs and legitimate desires of even a substantial fraction of the world's current (and continuously growing) numbers.

In my earlier volume I thus concluded with the suggestion that what the countries of the world had to strive for was what I had come to refer to as **'comprehensive human security'**. I did so especially in light of the ever more unsustainable conduct now in progress throughout most of the world—especially unsustainable as evidenced by the now almost inevitable global warming with the latter's dire consequences for both nature and humans (Westing 2013: 15–17).

So it is in this volume that I address the ever more evident shortcomings associated with traditional national security as I saw them, here selecting a number of illustrative publications of mine to do so. Thus it is that Chap. 3 covers my evolving understanding of the concept of security; Chaps. 4 and 6 deal with the need for a regional approach to security, with Chap. 7 offering the Korean peninsula as one case in point; Chap. 5 looks at the issue of security for the ocean, that huge extra-territorial region; Chap. 8 considers the pros and cons of our inexorably ongoing globalization as that phenomenon relates to security; and Chaps. 9 and 10 take a close look at overpopulation, perhaps the most flagrant of the stumbling blocks to achieving security. Preceding those eight Chapters just outlined, Chap. 2 provides a compilation of some three dozen or so additional articles of mine (both scholarly and popular) that appeared between 1983 and 2013 dealing with the issues surrounding the concept of security. I have done this in the hope that they could be of some use to those who might wish to delve more deeply into this critical field, that is, beyond those 11 articles reproduced in the present volume. Of particular interest might be the papers that were close contenders for reproduction in this volume (Westing 1981, 1989, 1991, 1992, 2001).

[2] Some additional early examinations of environmental security are also worth noting (Barnett 1984; Ehrlich and Ehrlich 1988; Gleick 1991; Myers 1986; Romulo et al. 1982; UNDDA 1986; UNEP 1988; Weinstein 1985).

1.2 What Next?

The necessary path forward is clear enough, but the obstacles to following it are immense. One small ray of hope is that at least some of the major powers have begun to minimize 'unnecessary' disruption of the environment during combat operations (Westing 2013: 137–146). On the other hand, we continue to pretty much witness business as usual despite the relentlessly increasing environmental and societal disruptions being brought about by ever greater levels of: (a) global warming; (b) global over-population; (c) global over-consumption; and (d) global forest destruction.[3] These unsustainable human actions must be reversed before it becomes too late to do so if we and our fellow creatures on earth are to survive and prosper, as we and they all should. Would that all the nations of the world not only adopt, but also faithfully adhere to, a code of action along the lines of *The Eritrean National Code of Conduct for Environmental Security* (cf. Appendix 1.1). That *Code*, after recalling especially the 1948 *Universal Declaration of Human Rights* and the 1982 *World Charter for Nature*, spells out in seven crucial steps what every nation in the world is at least morally obligated to follow in order to insure a decent long-term sustainable future for this earth and its divers inhabitants.

Thus, as I have suggested earlier (Westing 2013: 15), my hope is that pressure from an informed and sensitized public will lead to the necessary reorientation and restructuring of national priorities throughout the world in order to achieve the inexorably intertwined national environmental and social securities outlined above. But national restructuring will certainly not suffice without greater regional and global cooperation, to be achieved through a concomitant restructuring and strengthening of global governance.

Acknowledgments There are numerous people that it is my good fortune to be able to acknowledge for the assistance, encouragement, and advice that helped me so much in my work covered in this volume. In addition to those individuals singled out at the start of each of the Chapters 3 through 10, I must (as in my earlier volume) mention the following individuals to whom I owe very special debts of gratitude: (1) It was **Adele Smith Simmons**, my President at Hampshire College, who so importantly facilitated my studies. (2) It was **Frank Barnaby**, natural scientist and Director of SIPRI, **Sverre Lodgaard**, political scientist and Director of PRIO, **Naigzy Gebremedhin**, land planner and Head of Technology at UNEP, and **Mostafa K. Tolba**, natural scientist and Executive Director of UNEP, who (despite objections from the USA) all so unstintingly and liberally supported my SIPRI/PRIO/UNEP program and validated its results. (3) It was **Nicholas Polunin** (1909–1997), natural scientist and founding President of both the Foundation for Environmental Conservation and the World Council for the Biosphere (Westing 1997), as well as **Jozef Goldblat** (1923–2012), SIPRI's renowned expert on

[3] It is of some interest to note that the 2013 *Worldwide Threat Assessment of the US Intelligence Community* does, in fact, now recognize that 'competition and scarcity involving natural resources—food, water, minerals, and energy—are growing security threats' (p. 9), elaborating on that theme for several pages with the conclusion (p. 12) that the problem stems from a combination of climate change outside of historic norms and intense pressures from the world's growing population (cf. http://www.dni.gov/files/documents/Intelligence%20Reports/2013%20ATA%20SFR%20for%20SSCI%2012%20Mar%202013.pdf).

international law, who were such supportive colleagues, mentors, and friends (Westing 2012). (4) It is **Hans Günter Brauch**, long-time colleague and fellow peace researcher, for his unexpected and heart-warming initiative to have Springer Verlag honor me not only with the designation of *Pioneer on the Environmental Impact of War*, but additionally to have Springer ask me to prepare this follow-on volume. And finally (5), to repeat an earlier encomium, it is **Carol Eck Westing**, wife, close companion, gentle and insightful critic, to whom I owe the largest debt of gratitude for tolerating my eventful absences and, of course, for not standing in the way of my pursuit of these endeavors.

Appendix 1.1 The Eritrean National Code of Conduct for Environmental Security[4]

Appendix 1.1.1 The Government and Peoples of Eritrea

- Supportive of the 1948 Universal Declaration of Human Rights,[5] and affirming the 1981 African Charter of Human and Peoples' Rights[6];
- Recalling the 1972 Declaration on the Human Environment,[7] and conscious of the 1992 Declaration on Environment and Development;[8]
- In approval of the 1982 World Charter for Nature,[9] and in sympathy with the 1968 African Convention on the Conservation of Nature and Natural Resources,[10]
- Observant of the 1973 Convention on International Trade in Endangered Species of Wild Fauna and Flora,[11] in recognition of the 1992 Convention on Biological Diversity[12] and mindful of the 1944 Convention to Combat Desertification[13]; *,and*
- Aware that humankind not only depends upon the biosphere for its survival and well-being, but also that it must share that biosphere with the other living things on earth,

[4] Cf. this Chapter's unnumbered opening note.

[5] UN General Assembly Resolution No. 217(III)A, 10 December 1948.

[6] UNTS 26363.

[7] UN General Assembly Document No. A/CONF.48/14/Rev.1, pp. 3–5, November 1973.

[8] UN Document No. DPI/1344, pp. 9–11, April 1993.

[9] UN General Assembly Resolution No. 37/7, 28 October 1982.

[10] UNTS 14689.

[11] UNTS 14537.

[12] UNTS 30619.

[13] UNTS 33480.

Appendix 1.1.2 Herewith Solemnly Proclaim

1. A deep respect for all living things, and the natural environment upon which they depend, for each is a link in the chain that supports life on earth.
2. A firm endeavour to make use of the environment in such a fashion that no species will disappear as a result of domestic actions; *and*, in support of this endeavour, to maintain in perpetuity an adequate fraction of both the terrestrial and marine environments in their natural state; *and, further*, to eschew any trade in species of plants or animals threatened with extinction.
3. An unfailing dedication to maintain the national lithosphere (land), hydrosphere (water), and atmosphere (air) at levels of purity conducive to a healthy environment.
4. A steadfast resolve to utilise the national renewable natural resources sustainably and the non-renewable ones frugally, and also to dispose of all wastes sustainably; *and*, in support of this resolve, to achieve a national population level that is in balance with available national resources and sink capacities, so that both present and future generations can live in dignity, and especially so that development can be carried out sustainably and with equity.
5. A faithful desire to carry out no activity that would harm the environment beyond national boundaries.
6. A staunch commitment to cooperate as necessary with neighbouring states, and with the world community of nations, to protect and enhance the regional environment, the environment of regions beyond national jurisdiction, and the global biosphere in general; *and*, in support of this commitment, a constant devotion to resolve any environmental or other interstate dispute solely by amicable means.
7. A thorough acceptance of the need to infuse into all levels of the educational process social and environmental philosophies that would nurture an acceptance of the fundamental rights of both humans and nature.

References

Barnett, R.W. 1984. *Beyond War: Japan's Concept of Comprehensive National Security*. New York: Pergamon Press, 155 pp.

Behar, N. 1985. Non-military aspects of mutual security: regional and global issues. *Bulletin of Peace Proposals* [now *Security Dialogue*] (Oslo) 16:363–373.

Brown, L.R. 1977. *Redefining National Security*. Washington: Worldwatch Institute Paper No. 14, 46 pp.

Brundtland, G.H. et al. 1987. *Our Common Future: World Commission on Environment and Development*. rev edn. Oxford: Oxford University Press, 400 pp.

Ehrlich, P.R. & Ehrlich, A.H. 1988. *The Environmental Dimensions of National Security*. Stanford, CA: Stanford Institute for Population & Resource Studies, Paper No. 0013, 14 pp.

Gebremedhin, N. et al. 1989. *Environmental Security: a Report Contributing to the Concept of Comprehensive International Security.* Oslo: International Peace Research Institute Oslo, 24 pp.
Gleick, P.H. 1991. Environment and security: the clear connections. *Bulletin of the Atomic Scientists* (Chicago) 47(3):16–21.
IUCN. 1980. *World Conservation Strategy: Living Resource Conservation for Sustainable Development.* Gland, Switzerland: International Union for Conservation of Nature, portfolio (unpaginated).
Matthews, J.T. 1989. Redefining security. *Foreign Affairs* (New York) 68(2):162–177.
Myers, N. 1986. The environmental dimension of security issues. *The Environmentalist* (Lausanne, Switzerland) 6(4):251–257.
Romulo, C.P. et al. 1982. *Relationship between Disarmament and International Security.* New York: UN Department of Political & Security Council Affairs, Disarmament Study Series No. 8 (A/36/567), 55 pp.
Ullman, R.H. 1983–1984. Redefining security. *International Security* (Cambridge, MA) 8(1): 129–153.
UNDDA. 1986. *Concepts of Security.* New York: UN Department of Disarmament Affairs [now UN Office for Disarmament Affairs], Disarmament Study Series No. 14 (A/40/553), 53 pp.
UNEP. 1988. *Ad Hoc Experts' Meeting on Expanded Concept of International Security, Nairobi, 23–26 February 1988.* Nairobi: UN Environment Programme Document No. DOC.005E, 12 pp.
Weinstein, J.M. 1985. Nonmilitary threats to Soviet national security. *Naval War College Review* (Newport, RI) 38(4):28–40.
Westing, A.H. 1981. A world in balance. *Environmental Conservation* (Cambridge, UK) 8(3):177–183.
Westing, A.H. 1983. Pursuit of security: how long can the planet stand it? *Development Forum* (United Nations, Geneva) 11(3):1,10. April 1983.
Westing, A.H. 1986. Expanded concept of international security. In: Westing, A.H. (ed.). *Global Resources and International Conflict: Environmental Factors in Strategic Policy and Action.* Oxford: Oxford University Press, 280 pp: pp 183–200.
Westing, A.H. 1989. Environmental component of comprehensive security. *Bulletin of Peace Proposals* [now *Security Dialogue*] (Oslo) 20:(2):129–134.
Westing, A.H. 1991. We have to seek security in its broadest sense. *World Health Forum* (World Health Organization, Geneva) 12(2):137–139.
Westing, A.H. 1992. Environmental refugees: a growing category of displaced persons. *Environmental Conservation* (Cambridge, UK) 19(3):201–207.
Westing, A.H. 1997. Nicholas Polunin (1909–1997). *Independent* (London) 97(3483):18. 16 December 1997.
Westing, A.H. 2001. The environment, its natural resources, and international security: what are the connections?: what are the answers? *Environmental Awareness* (Vadodara, India) 24(2):45–51. April–June 2001.
Westing, A.H. 2012. *Jozef Goldblat (1923–2012): a professional and personal tribute.* Putney, VT: Westing Associates in Environment, Security, & Education, 4 pp. 30 September 2012.
Westing, A.H. 2013. *Arthur H. Westing: Pioneer on the Environmental Impact of War.* Heidelberg, Germany: Springer, 150 pp.

Chapter 2
The Author's Relevant Papers: A Selective Listing

*Note: The author of all the following entries is '**Westing, Arthur H.**' these having been extracted from his sequential life list of publications. The number preceding each title refers to its sequential number in that compilation. Publications by the author to which reference is made elsewhere in this text are keyed to that number. The publications below presented in this book are preceded by an asterisk (*).*

References

#116. A world in balance. *Environmental Conservation* (Cambridge, UK) 8(3):177–183. Autumn 1981.

#133. Pursuit of security: how long can the planet stand it? *Development Forum* (United Nations, Geneva) 11(3):1,10. April 1983.

#146. The arms race *versus* environmental and human welfare. In: UIP (ed.). *Ponències*. Sant Cugat del Vallès, Spain: Universitat Internacional de la Pau, 248 pp: pp 173–177. 1984. Reprinted in: *Man, Evolution, Cosmos* (Sofia) 3(2):50–53. 1984.

*#162. An expanded concept of international security. In: Westing, A.H. (ed.). *Global Resources and International Conflict: Environmental Factors in Strategic Policy and Action*. Oxford: Oxford University Press, 280 pp: pp 183–200 (Chap. 9). 1986. (***Nb:*** Reproduced herein as Chapter 3.).

#188. The environmental component of comprehensive security. *Bulletin of Peace Proposals* [now *Security Dialogue*] (Oslo) 20(2):129–134. June 1989.

*#191. Environmental approaches to regional security. In: Westing, A.H. (ed.). *Comprehensive Security for the Baltic: an Environmental Approach*. London: Sage Publications, 148 pp: pp 1–14 (Chap. 1). 1989. (***Nb:*** Major portions reproduced herein as part of Chapter 4.) Reprinted in German in: *Militärpolitik Dokumentation* (Frankfurt am Main) 14(83–84):5–16, 87–90. 1991.

*#192. Regional security in a wider context. In: Westing, A.H. (ed.). *Comprehensive Security for the Baltic: an Environmental Approach*. London: Sage Publications, 148 pp: pp 113–121 (Chap. 14). 1989. (***Nb:*** Major portions reproduced herein as part of Chapter 4.).

#193. *Comprehensive Security for the Baltic: an Environmental Approach*. (Editor & co-author). London: Sage Publications, xii + 148 pp. 1989.

#196. Comprehensive human security and ecological realities. *Environmental Conservation* (Cambridge, UK) 16(4):295. Winter 1989.

#197. Environmental security for the Danube basin. *Environmental Conservation* (Cambridge, UK) 16(4):323–329. Winter 1989.

A. H. Westing, *From Environmental to Comprehensive Security*, SpringerBriefs on Pioneers in Science and Practice 13,
DOI: 10.1007/978-3-319-00687-1_2,

*#203. Our place in nature: reflections on the global carrying-capacity for humans. In: Polunin, N. & Burnett, J.H. (eds). *Maintenance of the Biosphere*. Edinburgh: Edinburgh University Press, 228 pp: pp 109–120 (Chap. 8). 1990. (***Nb:*** Reproduced herein as Chapter 10.).

#207. The atmosphere as a common heritage of humankind: its role in environmental security. *Scientific World* (London) 34(4):5–6. 1990. (Also in French, German, & Russian editions) Earlier version in: *Environment* (Philadelphia, PA) 31(3):3–4. April 1989. Similar in: *Peace & the Sciences* [now *Peace & Security*] (Vienna) 1989(2–3):78–82. 1989. Similar in Norwegian in: *Harstad Tidende*, 12 February 1990, p. 22. Similar in Norwegian in: *Aftenposten* (Oslo) 131(73):2. 13 February 1990. Similar in Norwegian in: *Arbeiderbladet* (Oslo), 21 March 1990, p. 3. Similar in Norwegian in: *Friheten* (Oslo), 6 April 1990, p. 3. Later briefer version in: *Science* (Washington) 291(5505):827. 2 February 2001.

#210. We have to seek security in its broadest sense. *World Health Forum* (World Health Organization, Geneva) 12(2):137–139. 1991. (Also in Arabic, Chinese, French, Russian, & Spanish editions).

#213. Environmental security and its relation to Ethiopia and Sudan. *Ambio* (Stockholm) 20(5):168–171. August 1991.

#216. Disarmament, environment, and development and their relevance to the least developed countries: the military sector *vis-à-vis* security. In: Westing, A.H. (ed.). *Disarmament, Environment, and Development and their Relevance to the Least Developed Countries*. Geneva: UN Institute for Disarmament Research, Research Paper No. 10 (UNIDIR/91/83), 105 pp: pp 71–79 (Chap. 6). October 1991. Similar in: *Environmental Awareness* (Vadodara, India) 16(1):15–20. January–March 1993.

#226. Environmental refugees: a growing category of displaced persons. *Environmental Conservation* (Cambridge, UK) 19(3):201–207. Autumn 1992.

*#233. Environmental dimensions of maritime security. In: Goldblat, J. (ed.). *Maritime Security: the Building of Confidence*. UN Institute for Disarmament Research, Document No. UNIDIR/92/89, 159 pp: pp 91–102 (Chap. 6). 1992. (***Nb:*** Reproduced herein as Chapter 5.).

#236. Biodiversity and the challenge of national borders. *Environmental Conservation* (Cambridge, UK) 20(1):5–6. Spring 1993.

#237. Natural resources, conflict, and security, in a shrinking world. In: Nazim, M. & Polunin, N. (eds). *Environmental challenges: from Stockholm to Rio and Beyond*. Lahore: Energy & Environment Society of Pakistan & Foundation for Environmental Conservation, 284 pp: pp 85–116 (Chap. 4). 1993.

#239. Human rights and the environment. *Environmental Conservation* (Cambridge, UK) 20(2):99–100. Summer 1993. (Later revised version in: *Brattleboro [VT] Reformer* 96(298):4. 13 February 2009.).

#244. Environmental security for the Horn of Africa: an overview. In: Polunin, N. & Burnett, J.H. (eds). *Surviving with the Biosphere*. Edinburgh: Edinburgh University Press, 572 pp: pp 354–357. 1993.

*#245. Building confidence with transfrontier reserves: the global potential. In: Westing, A.H. (ed.). *Transfrontier Reserves for Peace and Nature: a Contribution to Human Security*. Nairobi: UN Environment Programme, 127 pp: pp 1–15 (Chap. 1). 1993. (***Nb:*** Reproduced herein as Chapter 6.) Shorter version in: *Environmental Awareness* (Vadodara, India) 18(1):11–22. January–March 1995.

#246. From hope to reality: establishing an Indochina tri-state reserve for peace and nature. In: Westing, A.H. (ed.). *Transfrontier Reserves for Peace and Nature: a Contribution to Human Security*. Nairobi: UN Environment Programme, 127 pp: pp 99–109 (Chap. 11). 1993.

#247. *Transfrontier Reserves for Peace and Nature: a Contribution to Human Security*. (Editor & co-author) Nairobi: UN Environment Programme, xi + 127 pp. 1993.

*#257. Population, desertification, and migration. *Environmental Conservation* (Cambridge, UK) 21(2):110–114,109. Summer 1994. (***Nb:*** Reproduced herein as Chapter 9.) Similar in: Puigdefábregas, J. & Mendizábal, T. (eds). *Desertification and Migrations*. Logroño, Spain: Geoforma Ediciones, 322 pp: pp 41–52. 1995.

#271. Environmental approaches to the avoidance of violent regional conflicts. In: Spillmann, K.R. & Bächler, G. (eds). *Environmental Crisis: Regional Conflict and Ways of Cooperation.* Zürich: Swiss Federal Institute of Technology, Center for Security Studies & Conflict Research & Swiss Peace Foundation, Environment & Conflicts Project, Occasional Paper No. 14, 162 pp: pp 148–153. September 1995. Shorter version in: *Environmental Awareness* (Vadodara, India) 20(1):19–24. January–March 1997.

#277. National security for Eritrea as a function of environmental security. *Environmental Awareness* (Vadodara, India) 18(3):83–87. July–September 1995.

*#282. Eritrean National Code of Conduct for Environmental Security. In: Gebremedhin, N. & Westing, A.H. (eds). *National Environmental Management Plan for Eritrea.* Asmara: Government of Eritrea, xii + 236 pp: pp xii, 229. January 1996. (***Nb:*** Reproduced herein as Chapter 1, Appendix 1.1.) Reprinted in: *Environmental Conservation* (Cambridge, UK) 22(1):78. Spring 1995. Reprinted in: *Environmental Awareness* (Vadodara, India) 18(3):87–88. July–September 1995. Reprinted in: *Earth Ethics* (Washington) 7(3–4):13. Spring–Summer 1996.

#287. The Eritrean–Yemeni conflict over the Hanish archipelago: toward a resolution favoring peace and nature. *Security Dialogue* (Oslo) 27(2):201–206. June 1996. Earlier version in: Häberli, R. *et al.* (eds.). *Proceedings of the Regional Workshop on Environmental Conflict Management and Sustainable Development, Asmara, Eritrea, 12–16 February 1996.* Zürich: Swiss Federal Institute of Technology, Center for Security Studies & Conflict Resolution, 289 pp: pp 240–257. April 1996.

#294. Canadian security in a broadened context. *Canadian Defence Quarterly* (Toronto) 26(2):13–14,16,18–19,22. Winter (December) 1996.

#306. Biodiversity losses in relation to national security. *Environmental Awareness* (Vadodara, India) 20(3):85–93. July–September 1997.

#309. Armed conflict and environmental security. *Environment & Security* (Sainte–Foy, Quebec) 1(2):31–39,11,21. 1997.

#318. Establishment and management of transfrontier reserves for conflict prevention and confidence building. *Environmental Conservation* (Cambridge, UK) 25(2):91–94. June 1998.

#328. Conflict *versus* cooperation in a regional setting: lessons from Eritrea. In: Suliman, M. (ed.) *Ecology, Politics and Violent Conflict.* London: Zed Books, 298 pp: pp 273–285 (Chap. 13). 1999.

#339. Biodiversity loss and its implications for security and armed conflict. In: Cracraft, J. & Grifo, F.T. (eds). *The Living Planet in Crisis: Biodiversity Science and Policy.* New York: Columbia University Press, 311 pp: pp 209–216 (Chap. 8). 1999.

#343. Towards regional environmental security. In: Gaan, N. (ed.). *Environment and National Security: the Case of South Asia.* New Delhi: South Asian Publishers, xii + 265 pp: pp v–ix. 2000.

*#347. Globalization *vis-à-vis* environmental security. *Environmental Awareness* (Vadodara, India) 23(1):7–15. January–March 2000. (***Nb:*** Reproduced herein as Chapter 8.).

#366. A Korean DMZ park for peace and nature: towards a code of conduct. In: Chae-Han Kim (ed.). *The Korean DMZ: Reverting Beyond Division.* Seoul: Sowha Publishing Company, 359 pp: pp 157–191 (Chap. 6). 2001. (Also published in a Korean-language edition in 2000 which included a Korean-language synopsis of this entry.) Earlier version in: IUCN (ed.). *Parks for Peace: International Conference on Transboundary Protected Areas as a Vehicle for International Co-operation, 16–18 September 1997, Somerset West, near Cape Town, South Africa: Conference Proceedings.* Gland, Switzerland: International Union for Conservation of Nature, 264 pp: pp 235–242. January 1998. Earlier version in: *International Environmental Affairs* (Hanover, NH) 10(1):8–17. Winter 1998 (1998–1999). Earlier version in: *Environmental Awareness* (Vadodara, India) 23(3):85–99. July–September 2000.

#371. The environment, its natural resources, and international security: what are the connections?: what are the answers? *Environmental Awareness* (Vadodara, India) 24(2):45–51. April–June 2001.

#375. Transfrontier reserves for peace and nature: saving two birds with one stone. In: Prakash, I. (ed.). *Ecology of Desert Environments: a Festschrift for Prof. J.L. Cloudsely–Thompson on his 80th birthday*. Jodhpur, India: Scientific Publishers, 471 pp: pp 23–32. 2001. An earlier synopsis in: *New York Times* 144(50114):A20. 6 July 1995.

#380. Peace, culture and ethics: recent history of conservation values in peace and war. In: Tolba, M.K. (ed.). 2001. *Our Fragile World: Challenges and Opportunities for Sustainable Development*. Oxford: EOLSS Publishers & Paris: UN Educational, Scientific & Cultural Organization, 2263 pp (2 vols): pp 865–872 (Vol. 1, Chap. 2.14). Earlier version in: *Environmental Conservation* (Cambridge, UK) 23(3):218–225. September 1996. Summary of earlier version in: Serageldin, I. & Barrett, R. (eds). *Ethics and Spiritual Values: Promoting Environmentally Sustainable Development*. Washington: International Bank for Reconstruction & Development (World Bank), Environmentally Sustainable Development Proceedings Series No. 12, 54 pp: p. 53. 1996.

#387. The business sector *vis-à-vis* environmental security. *Environmental Awareness* (Vadodara, India) 25(4):125–131. October–December 2002.

#432. Toward environmental sustainability and reduced tensions on the Korean peninsula. *Environment* (Philadelphia, PA) 52(1):20–23. January–February 2010. Summary in: *TBeNews* [newsletter of the IUCN (Gland, Switzerland) World Commission on Protected Areas, Transboundary Conservation Specialist Group: www.tbpa.net] 2010(1):5. April 2010.

*#440. All the many humans ever: an update. *BioScience* (Reston, VA) 60(10):777. November 2010. (***Nb:*** Reproduced herein as Chapter 10, Appendix 10.1.) Earlier version in: *BioScience* (Reston, VA) 31(7):523–524; 32(1)6. July–August 1981; January 1982.

*#441. *The Korean Demilitarized Zone (DMZ) as a Bridge between the two Koreas*. Berlin: Institute for Cultural Diplomacy, 25 pp. 11 November 2010. (***Nb:*** Reproduced herein as Chapter 7.).

#433. Population: the repressed dimension of global warming. *Brattleboro [VT] Reformer* 98(16):5. 20–21 March 2010. Similar in: *Environmental Awareness* (Vadodara, India) 32(1–2):13–14. January–June 2009. Earlier briefer version in: *International Herald Tribune* (Paris) 2010(39,482):6. 18 February 2010.

Part II
Benchmark Papers by the Author: A Selection

Chapter 3
From Environmental Security to Comprehensive Security: A Necessary Expansion

3.1 Introduction

The natural resources of the earth, both living and non-living, are most unevenly distributed. Many of them are thus in short supply or even absent in various countries. It is therefore not surprising that a substantial number of our many wars are fought over natural resources, whether or not this is made explicit by the initiating state (Westing 1986). Even when natural resources do not figure prominently as the cause of a war, they are often a contributing factor of some significance.

Demands on the land, fresh waters, and other natural resources of the earth are growing rapidly owing to the rapid increases in human numbers and to the even more rapid increases in human aspirations, the latter in both the developed and developing nations. This dilemma suggests that natural resources have the potential for playing an even more important role as a cause of war in the future than they have in the past. It is especially with this prospect in mind that two major issues are addressed in this Chapter: (a) means for reducing the likelihood of international conflict over natural resources; and (b) environmental measures for strengthening international security.

Military actions involving natural resources take one of two major forms: those the object of which is to obtain one or more natural resources and, conversely, those that would deny natural resources to an adversary. The former include wars of plunder, of aggrandizement, of colonial retention, and of dispute in contested or extra-territorial areas. The latter include wars or strategies of destruction, of attrition, and of siege or blockade. Means for preventing resource wars can be based on some combination of military posture, arms-control or disarmament

---

This Chapter is reproduced from the author's Entry #162 provided in Chapter 2, with the original title, 'An expanded concept of international security'. It is used here by permission of the Stockholm International Peace Research Institute, the copyright holder, as given on 13 March 2013. The author is pleased to acknowledge the able research assistance of J. Carol Barta.

A. H. Westing, *From Environmental to Comprehensive Security*,
SpringerBriefs on Pioneers in Science and Practice 13,
DOI: 10.1007/978-3-319-00687-1_3, © The Author(s) 2013

treaties, commercial (trade) agreements, domestic development, restraints on population growth, and restraints on the use of resources.

3.2 Towards the Prevention of Resource Wars

The natural resources over which wars might be waged can fall within the domain of a nation (territorial resources), can be international in the sense that they do not respect national boundaries (shared resources), or can be a part of the so-called international commons (extra-territorial resources). Very different problems are associated with these three categories, so that each deserves a separate section.

3.2.1 Territorial Resources

Wars that violate the integrity of another state, whether to gain possession of its natural resources or for some other aggressive end, are prohibited by international law and are today ostensibly rejected by the world at large. Indeed, some 60 nations formally condemned recourse to war in becoming states parties to the 1928 Pact of Paris (LNTS 2137) (Goldblat 1982: 136–137). Today, about 95 % of the 170 or so nations are members of the United Nations, which commits them to eschewing wars of aggression (UN 1968, Charter Article 2.4)[1]; and there exists an integral (albeit ineffectual) mechanism of conflict resolution (UN 1968, Charter Articles 33–51). Moreover, numerous bilateral and multilateral pacts of non–aggression or mutual assistance are in force.

Disputes over the location of a border between two states can be heightened, or even motivated, by the presence of natural resources in the contested strip, and such disagreements from time to time lead to armed conflict. The International Court of Justice provides one avenue for arbitration (UN 1968, Charter Articles 92–96).

In recent years a growing number of coastal states have laid exclusive claim to the natural resources of the ocean contiguous to their coastlines (Borgese and Ginsberg 1982: 564–568). Although these claims vary in width, many extend out to 370 km and thus encompass huge areas. These *de facto* claims have as yet no explicit basis in international law, yet appear to be generally recognized. An elaborate codification of the rules relating to such exclusive economic zones would be legally established were the 1982 Law of the Sea Convention to come into force.[2] Among other things, the Convention would provide for the conservation

[1] As at February 2013, there are perhaps 195 sovereign states (the 193 members of the United Nations plus Taiwan and Vatican City).

[2] The 1982 Law of the Sea Convention came into force in 1994 (UNTS 941116).

and optimum utilization of the living resources (Articles 61–62), for some level of 'equitable' participation by land–locked and geographically disadvantaged states in exploiting the 'surplus' of the living resources (Articles 69–71), and for regulations, enforcement, and the resolution of conflicts (various articles).

The 1982 Law of the Sea Convention is the product of a lengthy multinational effort under United Nations auspices. It would *inter alia* establish a comprehensive and integrated body of international law for this recent widespread extension of national jurisdiction over natural resources. Since formal adoption of the Convention is foundering because of the unpalatability to the USA and other advanced industrialized nations of other portions (cf. below), it may be useful to extract the portion devoted to exclusive economic zones for separate early adoption.

It is clear that the prevention of wars that aim at gaining control of territorial resources by plunder, aggrandizement, or similar means are subsumed under the overall efforts of the world community to preserve the sanctity of the sovereign state. Such interstate wars therefore require no further elaboration here. However, before leaving the matter of territorial resources it is important to mention two further matters: (a) colonial wars; and (b) domestic strife.

Numerous imperial colonies or similar possessions have attempted to secede from their metropolitan states and this has often led to intense warfare. The original acquisition of the possession—often a distant and at the time 'primitive' land—had in most instances been motivated by a desire for its land or other natural resources. And the reluctance to relinquish the possession has generally been based on similar natural–resource motives. Many of these armed attempts to gain independence met with failure in the early decades of this century, but in recent decades most have succeeded. Such wars of national liberation are now widely condoned and even recognized as permissible (UNGA 1974, Article 7); at any rate, relatively few colonies remain.

Political unrest within a country, rioting, and even *coups d'etat* in the Third World in recent years have been attributed to shortages of food (Wallensteen 1986). The implicated food deficits in Africa and elsewhere are the result of agricultural development not keeping up with rapidly growing populations, of reduced returns from nutrient–impoverished and eroded land being too heavily farmed or overgrazed or recklessly logged, and of other environmental, political, and social factors (Eckholm 1982; Sai 1984). The situation becomes especially acute from time to time in some of the more arid countries during the occasionally recurring years of particularly low rainfall. Many from among the excess rural populations—so-called environmental refugees[3]—migrate to urban centres, where living conditions become increasingly submarginal and chances for gainful employment are slim (El-Hinnawi 1985; Leroy 1986). The violent overthrow of the regime in Sudan in 1985 has been attributed in large measure to shortages and accompanying high prices of food (Fraser 1985). Approaches to alleviating these problems of domestic strife are addressed below.

---

[3] Cf. Chap. 6.

3.2.2 Shared Resources

Natural resources that do not respect national boundaries include fresh waters, ocean fisheries, and the atmosphere. The optimal and equitable utilization of such shared resources requires the establishment of a comprehensive body of law enjoying multilateral if not worldwide acceptance. Codification of this sort is also necessary to prevent disputes or to provide a vehicle for the peaceful resolution of those not prevented. No such codification is as yet in force despite various efforts, including recommendations by the United Nations Environment Programme (UNEP 1978a, b) and several attempts within the United Nations General Assembly over the past 25 years. Even simple resolutions establishing the principle of 'co-operation in the field of the environment concerning natural resources shared by two or more States' have received only weak support and none from the major powers (UNGA 1973). One very modest step towards protecting shared resources has been the establishment of transfrontier nature reserves (Thorsell 1985).[4]

The numerous bodies of fresh water (rivers, lakes, aquifers) that overlap national boundaries (cf. UN 1978) are a particular cause for future concern owing especially to the rapidly increasing agricultural and municipal demands for this fundamental resource (Falkenmark 1986). International upstream/downstream conflicts will arise both over allocation of use and over pollution control. Numerous disparate bilateral and regional water treaties exist, but these are often inadequate because they lack an integrated ecological or environmental approach and for other reasons (Biswas 1983; Caldwell 1984: 112–115, 308–309). An appropriate body of international law on shared bodies of fresh water is sorely needed, but remains an elusive goal.

Most ocean fish live over the continental shelves and thus largely within the rather newly proclaimed exclusive economic zones. Rational management of fish populations that overlap the boundaries of two or more such zones of national jurisdiction over natural resources, or the boundaries between such zones and the ocean beyond any national jurisdiction, requires international cooperation (Peterson and Teal 1986). Various multilateral treaties exist that deal with a particular species or class of fish (Kiss 1983; UNEP 1985), but adoption of the 1982 Law of the Sea Convention, or at least the relevant portion of it, would establish an appropriate body of comprehensive international law.

The atmosphere is shared by all the nations of the world and, owing to the air currents, constantly crosses national boundaries. Among a number of serious or potentially serious atmospheric problems (carbon dioxide enrichment, ozone depletion, air-lane crowding, weather modification, etc.), the one now most likely to lead to international conflict is that of noxious pollutants being introduced into the atmosphere by one country and blowing into and causing harm in another. There is usually little of a technical nature that a nation can do domestically to protect itself from foreign air pollution.

---

[4] Cf. Chap. 4.

International problems associated with air pollution are covered in part by a number of *ad hoc* bilateral and regional agreements as well as by one major multilateral instrument, the 1963 Treaty Banning Nuclear Weapon Tests in the Atmosphere, in Outer Space and Under Water (UNTS 6964) (Goldblat 1982: 157–158). This Partial Test Ban Treaty—a public–health and air–quality measure of some significance—is now adhered to by well over 100 nations, including the USA, the USSR, and the United Kingdom, but not China or France (neither of which, however, has carried out atmospheric tests in recent years). An important regional (European) agreement is the 1979 Convention on Long–Range Transboundary Air Pollution (UNTS 21623), which now has 24 or more states parties.

The atmosphere is another shared natural resource that calls for a comprehensive body of international law if future international conflict is to be avoided or dealt with in a reasonable fashion.

3.2.3 Extra-Territorial Resources

The ocean beyond any national jurisdiction plus the sea-bed beneath it, perhaps Antarctica plus the surrounding waters south of latitude 60°S, and the moon comprise the major extra-territorial domains that contain natural resources which either are exploitable or else may become so in the future. The upper atmosphere and outer space also deserve at least brief mention. The legal regimes pertaining to the various extra-territorial areas leave much to be desired, suggesting the possibility of future conflicts over the natural resources of these areas. The basic question is whether these areas or their resources can be laid claim to, and perhaps fought over, by individual nations or whether they are to be treated as a common heritage of humankind and managed accordingly for the equitably shared benefit of all.

The extra-territorial ocean, and the sea-bed beneath it, is distributed in three major and several minor basins that together represent more than half the global surface area (Westing 1980: 144–182). The extent of this immense public domain is perhaps 270 million km^2, the distance between sea-surface and sea-bed averages about 4 km, and its volume is thus of the order of 10^{18} m^3.

The extra-territorial ocean supports a modest population of fish which are widely dispersed and therefore not too readily exploitable (Peterson and Teal 1986). The sea-bed beneath the extra-territorial ocean is in various regions a treasure trove of minerals, including such strategic ones as manganese and cobalt (Cronan 1985; Stavridis 1985; Waldheim 1975). The depth of the sea-bed has to date precluded the commercial exploitation of these resources, but the time is not far off. More than 100 nations once declared support for the principle that the resources of the sea-bed beyond the limits of national jurisdiction be exploited for the benefit of humankind as a whole (UNGA 1970). The 1982 Law of the Sea Convention would establish this public domain as a common heritage of humankind, at the same time providing for its equitable management for the benefit of all. However, the advanced industrialized nations of the world—that is, those technically and economically capable of exploiting the resources of the deep sea-bed—have not been interested in supporting such a concept of an equitably shared international commons.

Antarctica is a huge, largely ice–covered, and essentially uninhabited continent the area of which has been for the most part (ca 85 %) divided up by the territorial claims (overlapping in three instances) of seven nations: Argentina, Australia, Chile, France, New Zealand, Norway, and the United Kingdom. Moreover, all of these claims have been contested by several major nations, including the USA and the USSR. The territorial disputes have been deferred for a time by the interested states parties through the vehicle of the 1959 Antarctic Treaty (UNTS 5778). This treaty *inter alia* demilitarizes the continent (Article 1), provides for the preservation of its living resources (Article 9.1.f), and permits its 'use' (Article 9.1.a). The terrestrial resources are not as yet commercially exploitable, but the associated fishery resources are being harvested by a number of nations within the limitations of two associated treaties: (a) the 1972 Convention for the Conservation of Antarctic Seals (UNTS 16529); and (b) the 1980 Convention on the Conservation of Antarctic Marine Living Resources (UNTS 22301). Despite several years of discussion among the states parties, no comparable associated treaty that would regulate the exploitation of the terrestrial and offshore mineral resources exists as yet.

It is clear that there is a potential for dispute and even armed conflict over the mineral resources of Antarctica, especially if existing or new territorial claims are exercised in the coming years (Kimball 1985; Luard 1983–1984; Shusterich 1984). The problem is exacerbated by the limited number of nations that support the treaty—only 32 at present—and the preferential treatment accorded to certain states parties—the 16 'consultative' (full) members.[5] In a sense, this group of full members has at least provisionally usurped the decision–making powers over Antarctica, in the process providing for the protection of the living natural resources of the continent, but at the same time not permitting it to become a common heritage of humankind.

The Svalbard archipelago in the Arctic Ocean deserves mention because it is in essence an extra-territorial area. The main island, Spitsbergen, has rich deposits of coal as well as lesser deposits of several other minerals, and its associated waters support a fishery resource. The 1920 Spitsbergen Treaty (LNTS 41) has established a legal regime for the archipelago. The treaty demilitarizes the area (Article 9) and provides for the preservation of the fauna and flora of the islands and associated waters (Article 2). All states parties to the treaty (40 at present[6]) enjoy access to the area and the right to exploit its natural resources 'on a footing of absolute equality', including specifically equal rights of fishing and hunting

[5] The 1959 Antarctic Treaty (UNTS 5778) has 50 states parties as at February 2013, of which 28 are 'consultative' (i.e., voting) members.

[6] The 1920 Spitsbergen Treaty (LNTS 41) has 41 states parties as at February 2013.

(Article 2) and of mining (Article 3). Norway—which was awarded limited sovereignty over the archipelago—was given the task, on behalf of the states parties, of establishing and overseeing the administrative regulations in support of the various stipulations of the treaty (Articles 2, 3, 8, 9, etc.). The Svalbard archipelago provides an early example of the management of an international commons, one in which the interested states parties (a non–exclusionary group) provide for the orderly utilization of the natural resources involved, but not for any sort of equitable sharing of the benefits derived from them.

The moon is the only celestial body that could conceivably become accessible for exploitation in the foreseeable future. At least two multilateral treaties would impinge upon such activity: (a) the 1967 Outer Space Treaty (UNTS 8843); and (b) the 1979 Moon Agreement (UNTS 23002). The Outer Space Treaty boasts more than 80 states parties,[7] including all of the major powers. The states parties have agreed that the moon 'is not subject to national appropriation' by any means (Article 2). It further commits the states parties to the 'use' of the moon 'for the benefit and in the interests of all countries, irrespective of their degree of economic or scientific development' and that the moon 'shall be the province of all mankind', for 'use by all States without discrimination of any kind' (Article 1). The treaty demilitarizes the moon (Article 4) and forbids 'harmful contamination' (Article 9).

The more recent 1979 Moon Agreement (UNTS 23002) repeats in essence all of the strictures quoted above from the 1967 Outer Space Treaty, but adds that 'due regard shall be paid to the interests of present and future generations as well as to the need to promote higher standards of living and conditions of economic and social progress and development' (Article 4) and also that measures be taken 'to prevent the disruption of the existing balance of its environment' (Article 7). However, in its most important innovation, the Agreement proclaims that 'the moon and its natural resources are the common heritage of mankind' and goes on to provide for the establishment of 'an international regime…to govern the exploitation of the natural resources of the moon', the purposes of which include 'an equitable sharing by all States Parties in the benefits derived from those resources' (Article 11). As is noted above with respect to the 1982 Law of the Sea Convention, the notion of a common heritage of humankind that is coupled with provisions for equitable sharing by all states parties appears to be unacceptable to any major power.[8] A legal analysis of the 'common heritage' principle with special reference to the moon can be found elsewhere (Halket et al. 1982–1983).

The upper atmosphere and outer space contain a number of extra-territorial resources of limited extent and thus subject to dispute. Of particular concern are (a) the communication portion of the electromagnetic spectrum (frequency range, ca 10×10^3–300×10^9 Hz) and (b) the geostationary (geosynchronous) satellite orbit (in the equatorial plane at an altitude of ca 36,000 km). The current status of

---

7 The 1967 Outer Space Treaty (UNTS 8843) has 100 states parties as at February 2013.

8 The 1979 Moon Agreement (UNTS 23002) has only 17 states parties as at February 2013.

these resources, how they are managed, and existing mechanisms for conflict resolution are discussed elsewhere (Jackson 1980; Smith 1983).

In conclusion, it is well known that extra-territorial resources are highly vulnerable to destructive exploitation unless they are managed within the framework of an enlightened, comprehensive system of international regulation. Such destruction has come to be referred to as the 'tragedy of the commons' (Hardin 1968; Hardin and Baden 1977). Thus, withholding support for such codification for the management of the natural resources of the ocean beyond any national jurisdiction or of the moon will not only prevent the establishment of effective mechanisms for conflict resolution, but will also lead to the extinction or exhaustion of the resources.

3.3 Towards the Establishment of a Secure Globe

The concept of national security is generally considered to refer to the security of a nation from being attacked by a potential enemy. States, of course, have the legitimate and necessary function—indeed, obligation—to protect their citizens from such a threat. And to this end most states maintain at least some sort of military establishment. However, it is not sufficient for states merely to ensure for everyone (to lean upon the *Universal Declaration of Human Rights*) 'life, liberty and security of person', but in addition to ensure for each citizen 'a standard of living adequate for the health and well–being of himself and of his family, including food, clothing [and] housing' and even 'a social and international order in which the rights…can be fully realized' (UNGA 1948, Articles 3, 25, 28).

There is, in fact, a growing recognition that threats to national and international security extend beyond external military threats (Behar 1985; Gordon 1978; Romulo et al. 1982; Thorsson et al. 1982; Ullman 1983–1984). Indeed, it has even been specifically recognized that such threats can arise from agricultural, natural–resource, and other environmental problems (Barnett 1984; Brown 1977; Purcell 1982; Weinstein 1985). The adverse effects of soil erosion, of water and air pollution, of harvesting renewable resources faster than their rates of renewal, of the rising demands for land, fresh water, fuels, and minerals, and of the accelerated rates of extinction of flora and fauna are all among the threats to national or international security in the expanded sense that have been identified as being of particular concern.

It is abundantly dear that many nations, regions, and the world as a whole are faced with serious environmental problems, various of which are becoming increasingly severe (Brown et al. 1985; Holdgate et al. 1982). Thus, any local, national, or multinational actions that come to grips with these problems—whether these be governmental, intergovernmental, or nongovernmental—will serve to reduce the threat to national and international security in the expanded sense. Moreover, it is clear that concerted actions of an international nature are often specifically called for because, as is noted, many of the problems are not confined to single countries, and also because some of the problems have progressed too far or are too complex to be dealt with by the intellectual or material resources of any one nation (IUCN 1980).

In considering the extent to which an amelioration of environmental problems would lead to a reduced threat of military action and to fewer wars, it is convenient to make a distinction between the living and non-living resources. There is, however, one principle that transcends this distinction: the concept that the earth and many of its most important natural resources constitute a common heritage of humankind that must be prudently and equitably managed for the benefit of all. If this concept of common heritage combined with shared benefits were embraced by the nations of the world, many potential threats to national and global security would become either moot or more readily soluble; and the likelihood of international conflict would be correspondingly reduced.

As to the living, renewable resources, the guiding principle must be, in the words of the *World Charier for Nature*, that such resources 'shall not be utilized in excess of their natural capacity for regeneration' (UNGA 1982, Principle II.10.d). In terms of human ecology, the carrying capacity must not be transgressed, even to the point that the 'sanctity of life' must give way before the 'sanctity of the carrying capacity' (Hardin 1985: 172). As stressed above, in order that this crucial aim be achieved without international conflict for those living resources that overlap national boundaries—and especially for those living resources that are found in international commons such as the extra-territorial ocean—agreed norms of conduct and associated codification are required, as embodied, for example, in the 1982 Law of the Sea Convention.

Within individual nations, especially within the many impoverished ones of the Third World, it is most important that population numbers (which are in some instances rising rapidly) be brought into balance on a long-term (sustained) basis with the capacities to produce food and wood (which are in some instances falling rapidly). In addition to the ecological and social or humanitarian benefits to be derived from such an achievement, that balance is likely to reduce domestic political turmoil and the number of violent changes of regime that occur so often in the developing countries. Thus, high priority must be given to instituting: (a) programmes of soil reclamation and improvement; (b) appropriate farm practices and technologies (including those of fish culture); (c) programmes of range management and forest management built around restoration and sustained yield of forage and wood; and (d) equitable distribution of, or access to, the nation's domestic land and water resources.

Among the non-living resources, the case of fresh waters has already been addressed in a prior section. Suffice it to say here that further conflict over shared fresh waters—regarding both equitable allocation and contaminants (water quality)—is imminent in various parts of the world in the absence of a relevant body of international law. And it could almost go without saying that nations must be able to bring into balance their demands for fresh water with the availability of that precious and often limited commodity.

For mineral fuels and non-fuel minerals, a strategy of sustained self-sufficiency is applicable only to the world as a whole. A large number of countries must depend upon external sources (i.e., on international trade) for many if not most of their oil and other mineral needs. It is thus evident that strategies for mineral security and the avoidance of conflicts over these resources are complex and at best not fully adequate (Arbatov 1986; Blechman 1985; Hveem 1986; Maull 1984; Solem and Scanlan 1986; Vogely 1982). Suggestions that can be made, most of them quite obvious, include: (a) the establishment of favourable international trade relations; (b) the parsimonious utilization of minerals not domestically available; (c) the use of (and continued research and development for) substitutes from available materials; and (d) the stockpiling of emergency supplies for use during temporary disruptions of supply. With respect to the energy resources, it has been shown, at least for the USA, that the present situation is vulnerable to disruption and disaster (Lovins and Lovins 1982). It was suggested that greater national security could be achieved by decentralization and greater reliance on a variety of non–renewable and renewable sources of energy, although not including nuclear energy.

It is perhaps fortunate for other states that the two superpowers are sufficiently well endowed with oil and, it seems, all other strategic minerals to be able to overcome interruptions of any foreign supply for extended periods of national emergency—preoccupied as they are with military might and the associated need for ready access to huge quantities of natural resources in time of war.

Reference was made earlier to the state's obligation, *inter alia*, to provide national military security. Any defensive military posture established by a state to that end must be as non–offensive and non–provocative as possible (Barnaby and Boeker 1982; Galtung 1982, 1984). One important component in providing such military security is for a nation to become as invulnerable as possible to enemy blockade or attack. Conforming to such a strategy provides one more motivation for a nation to become self-sufficient in the production of its staple foods, or potentially so. Moreover, such agricultural self-reliance should be independently achieved (or readily achievable) in as many subdivisions of the country as is feasible (Eide 1975).

3.4 Conclusion

Civilization is rooted in nature, an environment which has shaped human culture and influenced all scientific and artistic achievement. Indeed, humans are a part of nature, and life depends on the uninterrupted functioning of natural systems. It is thus inescapable that any concept of international security must in the last analysis be based on this obligate relationship of humankind with its environment. As has been recommended to the world community of nations by the United Nations Environment Programme (UNEP 1978b, Principle 1): 'It is necessary for States to co-operate in the field of the environment concerning the conservation and harmonious utilization of natural resources shared by two or more States. Accordingly, it is necessary that consistent with the concept of equitable utilization of shared natural resources, States co-operate with a view to controlling, preventing, reducing, or eliminating adverse environmental effects which may result from the utilization of such resources. Such co-operation is to take place on an equal footing and taking into account the sovereignty, rights, and interests of the States concerned.'

The global scarcity or uneven distribution of some of the natural resources upon which humans depend has often led to violent conflict. The attempt to eliminate competition over such resources as a source of international conflict must take several forms: (a) for the living natural resources (whether national, shared, or extraterritorial), an inviolate balance must be established between harvesting and natural regeneration; (b) for all natural resources that overlap national boundaries, formal mechanisms under international auspices must be established for equitable sharing; and (c) for all extra-territorial natural resources it must be accepted that they constitute a common heritage of humankind, and bodies of international law must be established accordingly that provide for their equitable sharing for the benefit of all.

The scarcity and degraded condition of crop lands, range (grazing) lands, and forest lands in many countries of the world has often led to serious medical and social problems, to political unrest, and to domestic violence and strife. In some developing countries these problems are of such magnitude and urgency that their alleviation and ultimate solution require the widest possible international cooperation and support. United Nations agencies might well be the avenue through which both material aid and the sharing of knowledge and skills could be channeled.

The attempt to eliminate such environmental calamities as a source of domestic, or even wider, conflict must again take several forms: (a) for the degraded soils, an urgent programme of rehabilitation must be established; (b) for the over-used grazing lands and forest lands, a rigorous management regime must be established that limits intensity of exploitation to a level which permits regeneration; and (c) a long-term balance must be achieved between the production of staple foods and other necessities of life, such as wood, on the one hand and the size of the population on the other—such balance to be achieved on a national and even local level.

The dimensions of the social and ecological problems facing humankind today are formidable, indeed. And the human record in dealing with major problems, whether real or perceived, does not provide much cause for optimism, as can be so amply demonstrated (e.g., Janis 1982; Tuchman 1984). But nevertheless, recognition of a problem is the first step to overcoming it. It is thus heartening to realize that at least 110 nations have been persuaded that (UNGA 1982): 'Competition for scarce resources creates conflicts, whereas the conservation of nature and natural resources contributes to justice and the maintenance of peace and cannot be achieved until [humankind] learns to live in peace and to forsake war and armaments'.

References

Arbatov, A.A. 1986. Oil as a factor in strategic policy and policy: past and present. In: Westing, A.H. (ed.). *Global Resources and International Conflict: Environmental Factors in Strategic Policy and Action*. Oxford: Oxford University Press, 280 pp: pp 21–37.

Barnaby, F. & Boeker, E. 1982. *Defence without Offence: Non-nuclear Defence for Europe*. Bradford, UK: Bradford University, School of Peace Studies Paper No. 8, 60 pp.

Barnett, R.W. 1984. *Beyond War: Japan's Concept of Comprehensive National Security*. New York: Pergamon Press, 155 pp.

Behar, N. 1985. Non-military aspects of mutual security: regional and global issues. *Bulletin of Peace Proposals* [now *Security Dialogue*] (Oslo) 16:363–373.

Biswas, A.K. 1983. Shared natural resources: future conflicts or peaceful development? In: Dupuy, R.-J. (ed.). *Settlement of Disputes on the New Natural Resources*. Hague: Martinus Nijhoff, 487 pp: pp 197–215.

Blechman, B.M. 1985. *National Security and Strategic Minerals: an Analysis of U.S. Dependence on Foreign Sources of Cobalt*. Boulder, Colorado: Westview Press, 96 pp.

Borgese, E.M. & Ginsburg, N. (eds). 1982. *Ocean Yearbook III*. Chicago: University of Chicago Press, 581 pp.

Brown, L.R. 1977. *Redefining National Security*. Washington: Worldwatch Institute Paper No. 14, 46 pp.

Brown, L.R. et al. 1985. *State of the World 1985*. New York: W.W. Norton, 301 pp.

Caldwell, L.K. 1984. *International Environmental Policy: Emergence and Dimensions*. Durham, NC: Duke University Press, 368 pp.

Cronan, D. 1985. Wealth of sea-floor minerals. *New Scientist* (London) 106(1459):34–38.

Eckholm, E.P. 1982. *Down to Earth: Environment and Human Needs*. New York: W.W. Norton, 238 pp.

Eide, A. 1975. Planting every inch. *Ceres* (Rome) 8(1):50–53.

El-Hinnawi, E. 1985. *Environmental Refugees*. Nairobi: UN Environment Programme, 41 pp.

Falkenmark, M. 1986. Fresh waters as a factor in strategic policy and policy. In: Westing, A.H. (ed.). *Global Resources and International Conflict: Environmental Factors in Strategic Policy and Action*. Oxford: Oxford University Press, 280 pp: pp 85–113.

Fraser, R. (ed.). 1985. Sudan. *Keesing's Contemporary Archives* (Harlow, UK) 31(7):33700–33707.

Galtung, J. 1982. *Environment, Development and Military Activity: Towards Alternative Security Doctrines*. Oslo: Universitetsforlaget, 143 pp.

Galtung, J. 1984. *There are Alternatives!: Four Roads to Peace and Security*. Nottingham, UK: Spokesman, 221 pp.

Goldblat, J. 1982. *Agreements for Arms Control: a Critical Survey*. London: Taylor & Francis, 387 pp.

Gordon, L. 1978. *International Stability and North-South Relations*. Muscatine, IA: Stanley Foundation Occasional Paper No. 17, 35 pp.

Halket, T.D. et al. 1982–1983. Report on the proposed agreement governing the activities of states on the moon and other celestial bodies. *Jurimetrics Journal* (Chicago) 23(3):259–277.

Hardin, G. 1968. Tragedy of the commons. *Science* (Washington) 162:1243–1248.

Hardin, G. 1985. Human ecology: the subversive, conservative science. *American Zoologist* (USA) 25:469–476.

Hardin, G. & Baden, J. (eds). 1977. *Managing the Commons*. San Francisco: W.H. Freeman, 294 pp.

Holdgate, M.W. et al. (eds). 1982. *World Environment 1972–1982: a Report by the United Nations Environment Programme*. Dublin: Tycooly, 637 pp.

Hveem, H. 1986. Minerals as a factor in strategic policy and policy. In: Westing, A.H. (ed.). *Global Resources and International Conflict: Environmental Factors in Strategic Policy and Action*. Oxford: Oxford University Press, 280 pp: pp 55–84.

IUCN. 1980. *World Conservation Strategy: Living Resource Conservation for Sustainable Development*. Gland, Switzerland: International Union for Conservation of Nature, portfolio (unpaginated).

Jackson, C.L. 1980. Allocation of the radio spectrum. *Scientific American* (New York) 242(2):30–35,130.

Janis, I.L. 1982. *Groupthink: Psychological Studies of Policy decisions and Fiascoes*. 2nd edn. Boston: Houghton Mifflin, 351 pp.

Kimball, L. 1985. Antarctica: testing the great experiment. *Environment* (Philadelphia, PA) 27(7):14-17, 26–30.

Kiss, A.C. (ed.). 1983. *Selected Multilateral Treaties in the Field of the Environment*. Nairobi: UN Environment Programme, Reference Series No. 3, 525 pp.

Leroy, M. 1986. Human population as a factor in strategic policy and policy. In: Westing, A.H. (ed.). *Global Resources and International Conflict: Environmental Factors in Strategic Policy and Action*. Oxford: Oxford University Press, 280 pp: pp 159–182.

Lovins, A.B. & Lovins, L.H. 1982. *Brittle Power: Energy Strategy for National Security*. Andover, MA: Brick House, 486 pp.

Luard, E. 1983–1984. Who owns the Antarctic? *Foreign Affairs* (New York) 62:1175–1193.

Maull, H.W. 1984. *Raw Materials, Energy and Western Security*. London: Macmillan, 413 pp.

Peterson, S.B. & Teal, J.M. 1986. Ocean fisheries as a factor in strategic policy and policy. In: Westing, A.H. (ed.). *Global Resources and International Conflict: Environmental Factors in Strategic Policy and Action*. Oxford: Oxford University Press, 280 pp: pp 114–142.

Purcell, A.H. 1982. *Resource Optimization and World Peace*. Muscatine, IA: Stanley Foundation Occasional Paper No. 30, 22 pp.

Romulo, C.P. et al. 1982. *Relationship between Disarmament and International Security*. New York: UN Department of Political & Security Council Affairs, Disarmament Study Series No. 8 (A/36/567), 55 pp.

Sai, F.T. 1984. Population factor in Africa's development dilemma. *Science* (Washington) 226:801–805.

Shusterich, K.M. 1984. Antarctic treaty system: history, substance, and speculation. *International Journal* (Toronto) 39:800–827.

Smith, D.D. 1983. Conflict resolution in outer space: international law and policy. In: Dupuy, R.-J. (ed.). *Settlement of Disputes on the New Natural Resources*. Hague: Martinus Nijhoff, 487 pp: pp 243–277.

Solem, E. & Scanlan, A.F.G. 1986. Oil and natural gas as factors in strategic policy and policy: a long-term view. In: Westing, A.H. (ed.). *Global Resources and International Conflict: Environmental Factors in Strategic Policy and Action*. Oxford: Oxford University Press, 280 pp: pp 38–54.

Stavridis, J. 1985. Resource wars. *U.S. Naval Institute Proceedings* (Annapolis. MD) 111(1):72–77.

Thorsell, J. 1985. Parks that promote peace. *WWF News* (Gland, Switzerland) 1985(38):2.

Thorsson, I. et al. 1982. *Relationship between Disarmament and Development*. New York: UN Department of Political & Security Council Affairs, Disarmament Study Series No. 5 (A/36/536), 189 pp.

Tuchman, B.W. 1984. *March of Folly: from Troy to Vietnam*. New York: Alfred A. Knopf, 447 pp+24 pl.

Ullman, R.H. 1983–1984. Redefining security. *International Security* (Cambridge, MA) 8(1):129–153.

UN. 1968. *Charter of the United Nations and Statute of the international Court of Justice*. New York: United Nations, 87 pp.

UN. 1978. Register of international rivers. *Water Supply & Management* (Oxford) 2(1):1-58+5 maps.

UNEP. 1978a. *Co-operation in the Field of the Environment concerning Natural Resources Shared by Two or More States*. Nairobi: UN Environment Programme, Decision No. 6/14, 24 May 1978, 1 p.

UNEP. 1978b. *Draft Principles of Conduct in the Field of the Environment for the Guidance of States in the Conservation and Harmonious Utilization of Natural Resources Shared by Two or More States*. Nairobi: UN Environment Programme, Document No. UNEP/GC.6/17, 10 March 1978, 3+14 pp.

UNEP. 1985. *Register of International Treaties and other Agreements in the Field of the Environment*. Nairobi: UN Environment Programme, Document No. UNEP/GC/INFORMATION/11/REV.1, 209 pp.

UNGA. 1948. *Universal Declaration of Human Rights*. New York: UN General Assembly Resolution No. 217(III)A, 10 December 1948, 3 pp.

UNGA. 1970. *Declaration of Principles Governing the Sea-bed and the Ocean Floor, and the Subsoil thereof, Beyond the Limits of National Jurisdiction*. New York: UN General Assembly Resolution No. 2749 (XXV), 17 December 1970, 2 pp.

UNGA. 1973. *Co-operation in the Field of the Environment concerning Natural Resources Shared by Two or More States*. New York: UN General Assembly Resolution No. 3129 (XXVIII), 13 December 1973, 1 p.

UNGA. 1974. *Definition of Aggression*. New York: UN General Assembly Resolution No. 3314 (XXIX), 14 December 1974, 6 pp.

UNGA. 1982. *World Charter for Nature*. New York: UN General Assembly Resolution No. 37/7, 28 October 1982, 5 pp.

Vogely, W.A. 1982. International security implications of materials and energy resource depletion. In: Teich, A.H. & Thornton, R. (eds). *Science, Technology, and the Issues of the Eighties: Policy Outlook*. Boulder, CO: Westview Press, 290 pp: pp 233–251.

Waldheim, K. 1975. *Economic Implications of Sea-bed Mining in the International Area*. New York: UN General Assembly Document No. A/CONF.62/37, 18 February 1975, 17 pp

Wallensteen, P. 1986. Food crops as a factor in strategic policy and policy: past and present. In: Westing, A.H. (ed.). *Global Resources and International Conflict: Environmental Factors in Strategic Policy and Action*. Oxford: Oxford University Press, 280 pp: pp 143–158.

Weinstein, J.M. 1985. Nonmilitary threats to Soviet national security. *Naval War College Review* (Newport, RI) 38(4):28–40.

Westing, A.H. 1980. *Warfare in a Fragile world: Military Impact on the Human Environment*. London: Taylor & Francis, 249 pp.

Westing, A.H. 1986. Environmental factors in strategic policy and action: an overview. In: Westing, A.H. (ed.). *Global Resources and International Conflict: Environmental Factors in Strategic Policy and Action*. Oxford: Oxford University Press, 280 pp: pp 3–20.

Chapter 4
Regional Security: An Ecological Necessity

4.1 Introduction

Most of the endless interstate conflicts that plague humankind are between nearby nations, and their ultimate resolution is often attempted by a threat of force if not an actual resort to arms. Some more or less modest fraction of these conflicts is at least in part over shared natural resources or derives in some other way from the human environment.

In many regions of the world the renewable natural resources are being utilized faster than their natural rate of renewal, and the regional environment is being abused in many other important ways as well. Moreover, the ecogeographical regions of the world are not generally delimited in synchrony with the political boundaries of the world. As a result, most environmental problems are shared by two or more of the 170 or so zealously sovereign states into which the peoples of the world have segregated themselves.[1] Coming to grips with most environmental problems in the face of this incongruence between environmental and political borders thus requires joint, or at least coordinated, action by nearby nations. Such interstate cooperation is, of course, in progress to a greater or lesser extent within many regions of the world.

This Chapter is reproduced from the author's Entries #191 and #192 provided in Chapter 2, with the original titles, 'Environmental approaches to regional security' and 'Regional security in a wider context', respectively; here combining major excerpts from each. They are used here by permission of the International Peace Research Institute Oslo and the United Nations Environment Programme, the joint copyright holders, each as given on 18 March 2013. The author is pleased to acknowledge the research assistance of T. Daniel Kjelling and Carol E. Westing.

---

[1] As at February 2013, there are perhaps 195 sovereign states (the 193 members of the United Nations plus Taiwan and Vatican City), of which 151 are coastal (including 49 which are islands) and 44 are landlocked.

A. H. Westing, *From Environmental to Comprehensive Security*,
SpringerBriefs on Pioneers in Science and Practice 13,
DOI: 10.1007/978-3-319-00687-1_4, © The Author(s) 2013

One question that can be addressed is the means by which environmental cooperation could be promoted among the states within a region, whether for the purpose of achieving greater comprehensive regional security or as an end in itself. However, another question of interest to address is the extent to which regional cooperation on environmental problems enhances regional cooperation in the diplomatic arena or on other fronts. In other words, to what extent does regional cooperation on environmental protection and natural-resource utilization serve as a confidence-building measure for the purpose of fostering comprehensive international security?

Dealt with first in this Chapter is the basic concept of environmental security (as well as some stumbling blocks to its achievement and maintenance); and second, what is meant by a region within the framework of the present study. Examined next is the idea of regional political security, and then that of regional environmental security. These considerations are followed by an initial exploration of how those two forms of security interact with each other, presumably leading to comprehensive regional security. Selected references to studies by others related to this overall subject are provided elsewhere (Westing 1989a).

4.2 The Concept of Environmental Security

Two major (though inextricably intertwined) prerequisites must be satisfied in order to be able to achieve environmental security (Westing 1986b, 1989b): (a) a *protection* requirement, that is, the quality of the human environment must be safeguarded; and (b) a *utilization* requirement, that is, any exploitation (harvesting or use) of renewable natural resources must be carried out on a sustaining basis. Thus, the management problems associated with environmental security fall into a number of more or less distinct categories and sub-categories:

1. Problems associated with *protection* of the environment: (a) avoidance of vandalism (wartime or other non-remunerative destruction); (b) avoidance of excessive pollution (air, water, soil), that is, pollution in excess of the natural renewal or cleansing processes; and (c) avoidance of any permanent anthropogenic intrusion whatsoever in a modest number of special areas; and
2. Problems associated with *utilization* of the environment: (a) avoidance of utilization at rates beyond long-term sustainability, that is, in excess of maximum sustained yield or maximum sustained discard; and (b) avoidance of utilization—in the event of past abuses—at rates that will prevent recovery of the degraded environment (recovery, moreover, that may well require human assistance).

The problems associated with environmental protection and utilization will, of course, vary in detail depending upon the nature of the resource. Resources can conveniently be divided into the following categories: (a) *non-extractive resources*, including the land and its soil (for agriculture, urbanization, transportation, waste disposal, etc.), water (for navigation, power, waste disposal, etc.), and the atmosphere (for communication, transportation, waste disposal, etc.); and (b) *extractive resources*, including non-renewable resources (mineral, etc.), and renewable resources (water [for drinking, irrigation, industry, etc.], wood, grass, fish, etc.).

It must be stressed here once again that all utilization of the renewable natural resources must be carried out on a sustained-yield basis; and that all disposal of wastes must be carried out on a sustained-discard basis. Indeed, without an inflexible commitment to the sustainable development of resources and the sustainable disposal of wastes there can be no environmental security. To lean upon one of the foundation stones of human ecology (Hardin 1985, p. 472): 'Thou shalt not transgress the carrying capacity'.

Before continuing, it will be useful to explain in somewhat greater detail some of the principal stumbling blocks to the achievement and maintenance of environmental security—namely, environmental vandalism, environmental pollution, habitat disruption, and resource over-utilization—and potentially available countermeasures.

4.2.1 Environmental Vandalism

Wanton disruption of the environment by armed conflict is a common occurrence in many ecogeographical regions of the world. The most important legal constraints on environmental vandalism are provided by 1977 Protocols I and II on the Protection of Victims of, respectively, International and Non-international Armed Conflict (UNTS 17512 and 17513) (Westing 1988b). Protocol I establishes that 'care shall be taken in warfare to protect the natural environment against widespread, long-term and severe damage' (Article 55.1). And both protocols prohibit attacks upon 'agricultural areas for the production of foodstuffs, crops, livestock, drinking water installations and supplies and irrigation works' (Articles 54.2 and 14, respectively) and upon 'works or installations containing dangerous forces, namely dams, dykes and nuclear electrical generating stations' (Articles 56.1 and 15, respectively). Additional contributions to the legal regime that would reduce wartime environmental vandalism are provided especially by the 1977 Environmental Modification Convention (UNTS 17119) (Westing 1984), the 1980 Conventional [Inhumane] Weapon Convention (UNTS 22495) (Westing 1988b).

To date, unconscionably low levels of support are enjoyed by Protocols I and II throughout the world. The achievement of environmental security cannot be assured without widespread adherence to these two instruments, both in letter and in spirit, this as a means of establishing and reinforcing appropriate environmental norms, both legal and cultural.

4.2.2 Environmental Pollution

Threats to environmental security from excess *air* pollution are reaching near-biospheric dimensions, to date still emanating largely from the developed world. The 1963 Partial Test Ban Treaty (UNTS 6964) prohibits transboundary radioactive air pollution (Westing 1988b). This is a most important treaty because it reinforces the notion that states do not have the right to degrade the environment of areas beyond the limits of their sovereignty. It is thus heartening to see that about two-thirds of all the world's nations are states parties to this treaty. However, the only significant regional air pollution treaty in the world applies only to Europe: the 1979 Convention on Long-range Transboundary Air Pollution (UNTS 21623) (Kiss 1983: 519–522). About three-quarters of all the European nations are states parties to this treaty. More generally, to lay the groundwork for establishing a sustained-discard and otherwise environmentally sound regime for the global atmosphere, it is clear that a comprehensive Law of the Air convention—comparable in principle to the 1982 Law of the Sea Convention (UNTS 31363)—is sorely needed (Westing 1989f).

Threats from excess *water* pollution are to date largely, but by no means entirely, a problem of ecogeographical regions in the developed world. The various regional seas regimes negotiated under the auspices of the United Nations Environment Programme (Nairobi) address the problem of water pollution (Gebremedhin 1989). Although a number of additional examples could readily be cited, most ecogeographical regions in the world have addressed this component of environmental security inadequately, if at all, whether in a legal or a practical sense. Thus the Baltic region *legal regime* should serve as a water-pollution-control model to the other ecogeographical regions elsewhere throughout much of the world (Broms 1989; Velner 1989; Westing 1989d).

4.2.3 Habitat Disruption

The very necessary strict protection of some small fraction of every ecogeographical region in the world from any human intrusion of a permanent nature is dealt with most inadequately in international law. However, it is at least facilitated by a number of multilateral treaties, for example, by the 1971 Convention on Wetlands of International Importance (UNTS 14583) and by the 1972 World Heritage Convention (UNTS 15511), providing, respectively, for the protection of internationally important wetlands especially as waterfowl habitat and of natural heritage sites of outstanding universal value.

At present, an overall 3 % of the world's national territories is reported to have a protected status as nature reserves, representing an amount that is at best one-half but probably only one-third the value it should be (Westing 1989e).[2] Moreover,

[2] Formally protected areas worldwide as of September 2012: 1.6 % of the ocean (although ca 7 % of territorial waters), and 12.7 % of the land (cf. http://www.unep-wcmc.org/ppr2012_903.html).

areas that have been particularly disrupted by urbanization, industrialization, or other human activity should be compensated for by the establishment and maintenance of nearby relatively undisturbed areas.

4.2.4 Resource Over-Utilization

The sustained and equitable utilization of the natural resources of most of the many ecogeographical regions of the world has yet to be achieved. By way of example, although the natural resources of the Baltic region (at least the living, renewable ones) are in principle well provided for (Westing 1989d), the realization of sustained-yield management in the region still remains somewhat elusive (Thurow 1989). A comparable example is provided by the Danube region (with its eight or so states), although to date an even less successful one in both legal and practical terms (Westing 1989c). A number of additional sustained-yield utilization regimes exist for other ecogeographical regions, at least on paper (e.g., Gebremedhin 1989). Nonetheless, on a worldwide basis the situation is bad and getting worse (Brown et al. 1989; UNEP 1987; WRI et al. 1988–1989). However, comfort can be drawn from the growing recognition of the problem within the international arena, both in its political sphere (Brundtland et al. 1987) and in its legal sphere (Singh 1988). The urgent need for 'sustainable development' is no longer an alien notion.

4.3 The Ecogeographical Region

The term 'region' is in the present study used to denote a geographical area that is unified in an ecological sense, gaining its integrity from this cohesion. The concept of such an *ecogeographical region* is essentially that of an ecological system, or ecosystem; that is to say, a unit made up of living and non-living components of the environment that interact to form a life-support system.

An ecogeographical region functions to some considerable extent independently of the regions contiguous to it, of more distant regions, and of the globe as a whole. However, it is by no means fully self-contained, having numerous more or less important links with the rest of the world. It is, of course, a part of the global ecosystem (the biosphere).

An ecogeographical region can be terrestrial or aquatic (freshwater or marine), or else include both terrestrial and aquatic components. The region can be inhabited or not; and it can subsume either or both of territorial (national) and extra-territorial (common-heritage) areas. If an ecogeographical region is inhabited, the criteria that contribute to its definition should be extended to include basic agricultural and silvicultural self-sufficiency.

It is never possible to define the boundaries of an ecogeographical region with enormous precision, although its approximate limits can often be established with reasonable confidence. Some clear examples of ecogeographical regions are seas with their associated drainage basins (watersheds; catchment areas), major rivers with their associated drainage basins, major mountain ranges, major islands or peninsulas, insular aggregations, deserts, tundras, and permanently ice-covered areas.

The various sorts of socially determined regions (ethnic ones, linguistic ones, political ones, and so forth) are, of course, superimposed upon the ecogeographical regions. These socially determined regions and their sub-regions are more or less difficult to define, and some are less stable and less effective than others (Brown 1986; Cantori and Spiegel 1970; Russett 1967; Väyrynen 1984). Moreover, they seldom coincide with their ecogeographical counterparts. However, it is precisely the lack of correspondence between ecogeographical regions and political regions, as well as the more ephemeral nature of the latter, that provide both the necessity for interstate cooperation and the challenge to bring it about.

4.4 Regional Political Security

There exists an understandable tendency to be swept up in the political security issues that preoccupy the two superpowers and thereby to dwell primarily upon the survival of humankind and related matters of far-reaching global security. However, most of the immediate political concerns of the numerous nations of the world are far more mundane, more localized, and perhaps even—it might be suggested—more tractable; hence the efficacy of a regional focus in a consideration of international relations.

Regional political security implies, first and foremost, the development of relations among the regional states sufficiently amicable and sufficiently codified to avoid armed conflict among them. The latter criterion—codification—can be achieved by a variety of formal mechanisms, that is, by the consummation of treaties that provide for submission to compulsory arbitration or similar obligate means of non-violent dispute resolution, that establish arms limitations, that dilute the existing national sovereignties through the creation of interstate federations of some sort or another, and so forth. However, if such instruments are to maintain the desired regional political security, the former criterion—amicability—must be satisfied as well.

More amicable regional political relationships can presumably be achieved by various means, both binding and non-binding—means now often referred to as confidence-building measures (Pfeiffer et al. 1982). This desired reduction of regional political tensions can be attempted by a host of unilateral (or multilateral) actions, among them: (a) renunciations of the use, testing, or even possession of unfailingly indiscriminate weapons (nuclear, biological, chemical, radiological, etc.); (b) shifts from traditional military postures (as these are reflected in both doctrine and capability) to more nearly defensive (non-provocative) ones; (c) overall reductions in military forces and expenditures; and (d) associated access to national territories adequate to permit the verification of the instituted actions.

The more amicable relationships being sought can presumably be fostered by a variety of supportive actions at national and sub-national levels (Dybern 1989; Westing 1988a). These supportive actions could be carried out, for example, by research institutes, schools, the mass media, professional associations, religious bodies, citizen groups, and even individuals. Moreover, the establishment or facilitation of regional interactions in the areas of commerce, science, sports, and tourism would all seem to be conducive to more amicable regional political relations—as presumably would cooperation in environmental protection and exploitation. It must not be overlooked that informal international scientific and other cooperation also makes a most valuable contribution to environmental security (Dybern 1989; Dybern and Pawlak 1989; Westing 1988c). The extent to which the 'high' politics of intergovernmental diplomacy is influenced by the 'low' politics of environmental and other technical cooperation is a major concern of the present study (cf., e.g., Vesa 1989).

At least six **justifications** can readily be mustered in support of an emphasis on regional political security. To begin with, despite the sophistication of today's systems of transportation and communication, the global community simply remains too large to be other than an abstraction for the vast majority of the more than 5,000 million people who comprise it at this moment.[3]

First: Regional efforts are more likely to succeed than global ones because the problems are more circumscribed and more clearly definable, the need for joint action is more readily apparent, and the potential partners—being tangible entities—are more easy to deal with;

Second: The endless interstate conflicts—a number of which involve the use of deadly force—are primarily between neighboring states. Hundreds of regional armed conflicts have occurred in this century alone, and at least several are always in progress in one region or another;

Third: The various regions of the world, being composed of different groupings of countries, differ sharply in their historical friendships and animosities, their aggregation of political systems, their ethnic compositions, their levels of development, and so forth. Since these and other national variables help to determine the nature of the regional security issues, the necessity for regionally distinct approaches becomes truly evident;

Fourth: Enhanced regional security serves to strengthen global security. Regional initiatives do so both directly and indirectly: directly by virtue of the region being a part of the whole; and indirectly by virtue of the region setting an example, not only for other regions, but also for the globe as a whole;

[3] The world population as at March 2013 is ca 7 billion, currently increasing at the compound rate of ca 1.1 % (giving a doubling time of ca 63 years). Cf. also Chap. 10, Appendix 10.1.

Fifth: A regional instrument would be useful in some cases even in the presence of a comparable global one. The latter might well be weaker than the former because the global common denominator is likely to be lower than the regional one; and

Sixth: Perhaps the most powerful justification for a regional approach to political security is that the ecogeographical region is endowed with an ecological integrity that most nations could not achieve short of resorting to conquest. The sacrosanct nature of political borders may well be a cornerstone of political philosophy and international relations. Nonetheless, the location of national borders is all too often arbitrary not only from the standpoint of ethnic, linguistic, or other demographic considerations, but, equally if not more important, from the standpoint of ecological considerations.

Numerous examples could be cited in empirical support of the concept of regional political cooperation (especially, of course, of bilateral cooperation). Two important binding legal precedents are in force at the worldwide level. In the 1945 United Nations Charter (now binding for 157 or more states[4]) it was agreed that: 'Nothing in the present Charter precludes the existence of regional arrangements or agencies for dealing with such matters relating to the maintenance of international peace and security as are appropriate for regional action…' (UN 1968, Article 52). And in the 1968 Nuclear-weapon Non-proliferation Treaty (UNTS 10485); now binding for 136 or more non-nuclear-weapon states) it was agreed that: 'Nothing in this Treaty affects the right of any group of States to conclude regional treaties in order to assure the total absence of nuclear weapons in their respective territories' (Goldblat 1982: 172–174, Article 7).

At least one important non-binding (recommendatory) precedent exists at the worldwide level. In 1971 the United Nations General Assembly solemnly declared that 'the Indian Ocean…together with the air space above and the ocean floor subjacent thereto, is hereby designated for all time as a zone of peace…' (in a resolution supported by 60 or more states) (UNGA 1971).

Various precedents for regional approaches to political security can be cited in which the primary participating states are from within the region itself (Willot et al. 1981). For example, two important regional treaties have as their major intent the realization of nuclear-weapon-free regions: (a) the 1967 Treaty for the Prohibition of Nuclear Weapons in Latin America (UNTS 9068) (Goldblat 1982:162–170); and (b) the 1985 South Pacific Nuclear Free Zone Treaty (UNTS 24592) (SIPRI 1986).

[4] The UN Charter (San Francisco, 26 June 1945; in force, 24 October 1945; UNTS unlisted) as at March 2013 is binding on 193 states.

4.5 Regional Environmental Security

Two major prerequisites must he satisfied in order to be able to achieve regional environmental security: (a) the quality of the human environment must be protected; and (b) any harvesting of renewable natural resources must be carried out on a sustaining basis (Westing 1989e).

Protecting the quality of the human environment implies the prevention of soil erosion, of air pollution, and of water pollution in excess of the natural renewal or cleansing processes; and also, of course, in excess of levels that would jeopardize the public health. It further implies the maintenance of representative habitats in their natural state and the prevention of species extinctions. In those instances where environmental damage or deterioration of some sort is already prevalent, protection of the human environment implies actions that would restore the damage, at least in so far as this remains possible.

The harvesting of renewable natural resources—primarily of wood, of grass (via livestock), and of fish—on a sustained-yield basis implies that the growing stock be maintained at a level such that the annually harvestable increment is maximized. Again, in those instances where a growing stock has been permitted to decline, actions are called for that would restore it to its optimal level.

The desirability of achieving regional environmental security, if not the actual necessity of doing so, is for the most part readily evident to anyone who has given the matter serious attention. And it is almost always the case that one or both of the two foundations of regional environmental security—the continuing maintenance of both environmental quality and quantity—require for their full implementation the joint action of all or most of the nations within the region. Again, the need for such regional cooperation is generally quite obvious.

A considerable number of instances of regional environmental cooperation could be cited. However, it should suffice to present 15 post-World War II examples in order to illustrate the environmental variety and distributional range of the sundry formal arrangements that exist and are in force (Kiss 1983; Lyster 1985; UNEP 1985–1987):

1. The 1949 Convention for the Establishment of an Inter-American Tropical Tuna Commission (UNTS 1041) is meant to maintain populations of yellowfin tuna (*Neothunnus macropterus* = *N. albacares*; IUCN Near Threatened) and skipjack tuna (*Katsuwonus pelamis*; IUCN Least Concern) in the eastern Pacific Ocean in order to permit maximum sustained catches.
2. The 1952 International Convention for the High Seas Fisheries of the North Pacific Ocean (UNTS 2770) is meant to ensure maximum sustained productivity of the fishery resources of that region, and also to coordinate regional research and conservation.
3. The 1956 Plant Protection Agreement for the Asia and Pacific Region (UNTS 1963) is meant to prevent the introduction into that region, and the spread within it, of destructive plant diseases and pests.

4. The 1958 Convention concerning Fishing in the Waters of the Danube (UNTS 4845) is meant to achieve the rational utilization and expansion of the fish stocks in that river system, *inter alia*, by improving the natural conditions for fish breeding and by preventing pollution.
5. The 1959 Convention concerning Fishing in the Black Sea (UNTS 5402) is meant to ensure the rational utilization and development of the fishery resources in that Sea.
6. The 1963 Agreement concerning the International Commission for the Protection of the Rhine against Pollution (UNTS 23469) is meant to maintain cooperation among the riparian states in preventing pollution and in improving the quality of that river system.
7. The 1973 Convention on Fishing and Conservation of the Living Resources in the Baltic Sea and Belts (UNTS 16710) is meant to provide for the cooperative management of all of the Sea's renewable natural resources; and which, *inter alia*, established an International Baltic Sea Fishery Commission (Warsaw) (Westing 1989d).
8. The 1973 Agreement on Conservation of Polar Bears (Oslo, 15 November 1973; in force, 26 May 1976; UNTS unlisted) is meant to protect the polar bear (*Ursus maritimus* = *Thalarctos maritimus*; IUCN Vulnerable) in the Arctic region through conservation and management measures, *inter alia,* through hunting restrictions and through actions to preserve the ecosystem of which the bears are a part.
9. The 1974 Convention on the Protection of the Marine Environment of the Baltic Sea Area (UNTS 25986) is meant to protect the Baltic environment from all forms of degradation; and which, *inter alia*, established a Baltic Marine Environment Protection Commission (Helsinki) (Velner 1989).[5]
10. The 1976 Convention for the Protection of the Mediterranean Sea against Pollution (UNTS 16908) is meant to achieve a coordinated and comprehensive approach to the protection and enhancement of the Mediterranean region.
11. The 1978 Kuwait Regional Convention for Co-operation on the Protection of the Marine Environment from Pollution (UNTS 17898) is meant to prevent, abate, and combat pollution in the Persian Gulf.
12. The 1978 Treaty for Amazonian Co-operation (UNTS 19194) is meant to permit the harmonious and mutually beneficial development of the Amazon region and to achieve the preservation of the environment of this major river system as well as the conservation and rational utilization of its natural resources.
13. The 1979 Convention on the Conservation of European Wildlife and Natural Habitats (UNTS 21159) is meant to conserve especially those endangered species of wild flora and fauna, and their natural habitats, the conservation of which requires interstate cooperation.

[5] The 1974 Convention on the Protection of the Marine Environment of the Baltic Sea Area (UNTS 25986) was superceded by the 1992 Convention of same name (UNTS 36495).

14. The 1982 Regional Convention for the Conservation of the Red Sea and Gulf of Aden Environment (Jiddah, Saudi Arabia, 14 February 1982; in force, 20 August 1985; UNTS unlisted) is meant to ensure rational use of the living and non-living marine and coastal resources of that region.
15. The 1983 Convention for the Protection and Development of the Marine Environment of the Wider Caribbean Region (UNTS 25974) is meant to protect and manage that region with its associated coastal areas, *inter alia*, by controlling pollution from all sources and by protecting or preserving rare or fragile ecosystems as well as the habitat of endangered species.

The selection of environmental treaties just outlined suggests the clarity with which the international community recognizes the significance of the ecogeographical region (seas, river systems, insular aggregations, etc.) as a unifying concept, and also the importance of protecting and developing it on a sustainable basis. Even more important, the treaties demonstrate the expanding recognition of a need for multilateral cooperation. Finally, an analysis of the four regional agreements enumerated above that were concluded within the framework of the Regional Seas Programme of the United Nations Environment Programme (Nairobi) makes it clear that the prerequisites for regional success in these matters include (Gebremedhin 1989): (a) a carefully coordinated process of preparation; (b) full or virtually full regional participation; and (c) a continuing coordinating agency that is regionally perceived as being neutral and trustworthy.

4.6 Comprehensive Regional Security

Comprehensive regional security demands both the prevention of armed conflict and the fulfillment of basic human needs and amenities. It is clear that environmental factors play an important role in satisfying the former of these two requirements and a key role in the latter (Westing 1986b).

With respect to regional conflict, some level of regional political insecurity can be attributed to the export of pernicious wastes by transboundary air or water currents. Further regional political insecurity can be attributed to the disproportionate exploitation of natural resources being shared by two or more states, or by the overall depletion of such resources. Various wars can even be attributed to the desire by a nation to allay its natural-resource insufficiencies (Westing 1986c). From considerations such as these it follows that regional political security will be strengthened through curtailment of transboundary pollution, through maximization of the sustained yield of shared renewable natural resources, and through establishment of equitable means of allocating shared (and common-heritage) natural resources.

Having earlier established the basis, prevalence, and importance of regional cooperation that focuses on political security, and having accomplished the same for regional cooperation that focuses on environmental security, it now remains to explore the more or less subtle interactions between these two domains. Thus it would be of interest to learn whether progress in the one facilitates significant progress in the other; or conversely, whether reversals in the one lead to significant reversals in the other.

To begin with, it is clear that regional liaisons among states established for political or economic purposes facilitate their subsequent cooperation on environmental matters. For example, such a quintessentially political union as the North Atlantic Treaty Organization (Brussels), founded in 1949, in 1969 added environmental concerns to its agenda, when its newly established Committee on Challenges of Modem Society began, *inter alia*, studying environmental pollution and the rational use of energy (NATO 1985: 29–30). And the Council for Mutual Economic Assistance (Moscow), also founded in 1949, in 1973 initiated a broad range of cooperative environmental projects among its members (CMEA 1979: 30–53; cf. also Gorbachev 1987).

Conversely, strong political tensions within a region can readily hinder the establishment of environmental cooperation. For example, it seems clear that consummation of the 1973 Baltic Fishing Convention (UNTS 16710) and the 1974 Baltic Marine Environment Convention (UNTS 25986) had to await recognition in 1973 by the Federal Republic of Germany and some of the other Baltic littoral states of the legitimacy of the German Democratic Republic (Rytövuori 1980; Vesa 1989).

On the other hand, political animosities within a region need not exclude the possibility of at least some level of environmental cooperation. Thus, the nations united by the Mekong River system have been working together to a greater or lesser extent on water management during most of the past three decades despite the Second Indochina war of 1961–1975 and other regional hostilities since its termination (Caldwell 1984: 114; Sewell and White 1966). At least two of the regional environmental treaties referred to earlier include some parties which are more or less strongly antagonistic towards each other: (a) the 1976 Convention for the Protection of the Mediterranean Sea (UNTS 16908) includes among its partners Israel on the one hand and Libya and Syria on the other; and (b) the 1978 Kuwait Regional Convention for Co-operation on the Protection of the Marine Environment from Pollution [in the Persian Gulf] (UNTS 17898) includes as partners both Iran and Iraq. However, whether these formal environmentally based liaisons have any importance regarding the adversarial political relationships among these states is not evident.

One indirect contribution of environmental security to comprehensive security is via the alleviation of some of the causes of military insecurity, to wit those brought about by resource scarcities, whether national, regional, or overall (Westing 1986a, c).

A second indirect contribution of environmental security to comprehensive security is via its international confidence-building effect, as envisioned by the 1975 Final Act on Security and Co-operation in Europe (Helsinki, 1 August 1975). Indeed, a host of confidence-building measures can be envisioned within military, economic, technological, environmental, and other spheres (Lodgaard 1989;

Pfeiffer et al. 1982)—each one useful not only for building confidence, but as well in its own right. As one possible example of successful confidence-building, it has been suggested that economic cooperation among the Balkan states was a precursor to warmer political relations within that region (Behar 1985). However, the influence of confidence-building measures on intergovernmental relations—that is, the influence of actions falling within the 'functionalist' realm of 'low' politics on actions falling within the 'international regime' realm of 'high' politics—appears in general to be quite subtle and rather slow (Domke 1988: 140–156; Vesa 1989). It is perhaps safe to suggest that neither the 'functionalist' approach (Groom 1975) nor the 'international regime' approach (Mingst 1981) suffices to achieve the 'working peace system' of comprehensive security, both being necessary, each reinforcing the other.

4.7 Conclusion

'Security' is a legitimate quest for all. And it has been quite evident for some time now that it has been necessary to expand the concept of security far beyond the traditional notion of military security. The first step in this evolution was to incorporate into the concept of security the notion of human rights, as exemplified, for example, by the Universal Declaration of Human Rights, which proclaims that (UNGA 1948): 'everyone has the right to life, liberty and the security of person' (Article 3), 'to a standard of living adequate for…health and well-being…including food, clothing, [and] housing' (Article 25), and 'to a social and international order in which [these] rights and freedoms…can be fully realized' (Article 28).

More recently it has become painfully clear that the social security proclaimed by the Universal Declaration of Human Rights is unachievable unless it is treated within a framework of security that is expanded once again. Thus, the next step in the evolution of the concept of security—and perhaps the most important one—has been to include the notion of environmental constraints, as exemplified, for example, by the World Charter for Nature, which proclaims that (UNGA 1982): 'nature shall be respected and its essential processes…not…impaired' (Article 1), 'living resources shall not be utilized in excess of their natural capacity for regeneration, the productivity of soils shall be maintained or enhanced, [and] non-renewable resources…shall be exploited with restraint' (Article 10); moreover, 'nature shall be secured against degradation caused by warfare or other hostile activities' (Article 5) and 'military activities damaging to nature shall be avoided' (Article 20). It is thus a good sign that the field of international environmental diplomacy is burgeoning (Caldwell 1984; Carroll 1988).

With the existence of an adequately expanded concept of 'security', the time has become ripe to address the various associated problems in a concerted manner. Doing so at the regional level is extremely important in its own right—and seems possible. The strengthening of regional security will also go a long way towards making the world more secure. Global security will be enhanced because each region is a part of the whole. It will be further enhanced because regional cooperation will facilitate the worldwide cooperation necessary to tackle the several major problems having truly global dimensions.

Now it remains to be seen whether the nations of the world will rise in time to the formidable challenge of relaxing their outdated notions of sovereignty, thereby enabling them to achieve comprehensive international security, first regionally and, in time, globally.

References

Behar, N. 1985. Non-military aspects of mutual security: regional and global issues *Bulletin of Peace Proposals* [now *Security Dialogue*] (Oslo) 16:363–373.

Broms, B. 1989. Multilateral agreements in the Baltic region. In: Westing, A.H. (ed.). *Comprehensive Security for the Baltic: an Environmental Approach.* London: Sage Publications, 148 pp: pp 62–71.

Brown, L.H. 1986. Regional collaboration in resolving third-world conflicts. *Survival* (London) 28:208–220.

Brown, L.R. et al. 1989. *State of the World 1989: a Worldwatch Institute Report on Progress Toward a Sustainable Society*. New York: W.W. Norton, 256 pp.

Brundtland, G.H. et al. 1987. *Our Common Future: World Commission on Environment and Development*. rev edn. Oxford: Oxford University Press, 400 pp.

Caldwell, L.K. 1984. *International Environmental Policy: Emergence and Dimensions*. Durham, NC: Duke University Press, 368 pp.

Cantori, L.J. & Spiegel, S.L. (eds). 1970. *International Politics of Regions: a Comparative Approach*. Englewood Cliffs, NJ: Prentice-Hall, 432 pp.

Carroll, J.E. (ed.). 1988. *International Environmental Diplomacy: the Management and Resolution of Transfrontier Environmental Problems.* Cambridge, UK: Cambridge University Press, 291 pp.

CMEA. 1979. *Environmental Protection in CMEA Member Countries*. Moscow: Council for Mutual Economic Assistance, 270 pp.

Domke, W. K. 1988. *War and the Changing Global System*. New Haven: Yale University Press, 209 pp.

Dybern, B.I. 1989. Nongovernmental scientific organizations relevant to the Baltic region. In: Westing, A.H. (ed.). *Comprehensive Security for the Baltic: an Environmental Approach.* London: Sage Publications, 148 pp: pp 87–89.

Dybern, B.I. & Pawlak, J.F. 1989. International Council for the Exploration of the Sea. In: Westing, A.H. (ed.). *Comprehensive Security for the Baltic: an Environmental Approach.* London: Sage Publications, 148 pp: pp 81–86.

Gebremedhin, N. 1989. Lessons from the UNEP Regional Seas Programme. In: Westing, A.H. (ed.). *Comprehensive Security for the Baltic: an Environmental Approach.* London: Sage Publications, 148 pp: pp 90–98.

Goldblat, J. 1982. *Agreements for Arms Control: a Critical Survey*. London: Taylor & Francis, 387 pp.

Gorbachev, M. 1987. Reality and guarantees for a secure world. *New Times* (Moscow) 39(3):3–6.

Groom, A.J.R. 1975. Functionalist approach and East/West cooperation in Europe. *Journal of Common Market Studies* (London) 13:21–60.

Hardin, G. 1985. Human ecology: the subversive, conservative science. *American Zoologist* (USA) 25:469–476.

Kiss, A.C. (ed.). 1983. *Selected Multilateral Treaties in the Field of the Environment*. Nairobi: UN Environment Programme, Reference Series No. 3, 525 pp.

Lodgaard, S. 1989. Confidence building in the Baltic region. In: Westing, A.H. (ed.). *Comprehensive Security for the Baltic: an Environmental Approach*. London: Sage Publications, 148 pp: pp 99–112.

Lyster, S. 1985. *International Wildlife Law: an Analysis of International Treaties concerned with the Conservation of Wildlife*. Cambridge, UK: Grotius, 470 pp.

Mingst, K.A. 1981. Functionalist and regime perspectives: the case of Rhine River cooperation. *Journal of Common Market Studies* (London) 20:161–173.

NATO. 1985. *NATO Handbook*. Brussels: North Atlantic Treaty Organization, 105 pp.

Pfeiffer, G. et al. 1982. *Comprehensive Study on Confidence-building Measures*. New York: UN General Assembly, Centre for Disarmament, Disarmament Study Series No. 7 (A/36/474), 42 pp.

Russett, B.M. 1967. *International Regions and the International System: a Study in Political Ecology*. Chicago: Rand-McNally, 252 pp.

Rytövuori, H. 1980. Structures of détente and ecological interdependence: cooperation in the Baltic Sea area for the protection of marine environment and living resources. *Cooperation & Conflict* (Oslo) 15:85–102.

Sewell, W.R.D. & White, G.F. 1966. Lower Mekong: an experiment in international river development. *International Conciliation* (New York) 1966(558):1–63.

Singh, N. 1988. Sustainable development as a principle of international law. In: Waart, P. de et al. (eds). *International Law and Development*. Dordrecht: Martinus Nijhoff, 457 pp: pp 1–12, 419–424.

SIPRI. 1986. South Pacific Nuclear Free Zone Treaty. *SIPRI Yearbook* (London) 1986:509–519.

Thurow, F. 1989. Fishery resources of the Baltic region. In: Westing, A.H. (ed.). *Comprehensive Security for the Baltic: an Environmental Approach*. London: Sage Publications, 148 pp: pp 54–61.

UNEP. 1985–1987. *Register of International Treaties and other Agreements in the Field of the Environment*. Nairobi: UN Environment Programme, Document No. UNEP/GC/INFORMATION/11/REV.1+Supplement 1, 209+46 pp.

UNEP. 1987. *State of the World Environment 1987*. Nairobi: UN Environment Programme, Publication No. UNEP/GC.14/6, 76 pp.

UNGA. 1948. *Universal Declaration of Human Rights*. New York: UN General Assembly Resolution No. 217(III)A, 10 December 1948, 3 pp.

UNGA. 1971. *Declaration of the Indian Ocean as a Zone of Peace*. New York: UN General Assembly Resolution No. 2832(XXVI), 16 December 1971, 2 pp.

UNGA. 1982. *World Charter for Nature*. New York: UN General Assembly Resolution No. 37/7, 28 October 1982, 5 pp.

Väyrynen, R. 1984. Regional conflict formations: an intractable problem of international relations. *Journal of Peace Research* (Oslo) 21:337–359.

Velner, H.A. 1989. Baltic Marine Environment Protection Commission. In: Westing, A.H. (ed.). *Comprehensive Security for the Baltic: an Environmental Approach*. London: Sage Publications, 148 pp: pp 75–80.

Vesa, U.J. 1989. Political security in the Baltic region. In: Westing, A.H. (ed.). *Comprehensive Security for the Baltic: an Environmental Approach*. London: Sage Publications, 148 pp: pp 35–45.

Westing, A.H. (ed.). 1984. *Environmental Warfare: a Technical, Legal and Policy Appraisal*. London: Taylor & Francis, 107 pp.

Westing, A.H. 1986a. Environmental factors in strategic policy and action: an overview. In: Westing. A.H. (ed.). *Global resources and international conflict: environmental factors in strategic policy and action*. Oxford: Oxford University Press, 280 pp: pp 3–20.

Westing, A.H. 1986b. Expanded concept of international security. In: Westing, A.H. (ed.). *Global Resources and International Conflict: Environmental Factors in Strategic Policy and Action*. Oxford: Oxford University Press, 280 pp: pp 183–200.

Westing, A.H. 1986c. Wars and skirmishes involving natural resources: a selection from the twentieth century. In: Westing, A.H. (ed.). *Global Resources and International Conflict: Environmental Factors in Strategic Policy and Action*. Oxford: Oxford University Press, 280 pp: pp 204–210.

Westing, A.H. (ed.). 1988a. *Cultural Norms, War and the Environment*. Oxford: Oxford University Press, 177 pp.

Westing, A.H. 1988b. Multilateral treaties constraining military disruption of the environment: excerpts. In: Westing. A.H. (ed.). *Cultural Norms, War and the Environment*. Oxford: Oxford University Press, 177 pp: pp 163–168.

Westing, A.H. 1988c. Towards non-violent conflict resolution and environmental protection: a synthesis. In: Westing. A.H. (ed.). *Cultural Norms, War and the Environment*. Oxford: Oxford University Press, 177 pp: pp 151–159.

Westing, A.H. 1989a. Environmental approaches to regional security: select bibliography. In: Westing, A.H. (ed.). *Comprehensive Security for the Baltic: an Environmental Approach*. London: Sage Publications, 148 pp: pp 122–123.

Westing, A.H. 1989b. Environmental component of comprehensive security. *Bulletin of Peace Proposals* [now *Security Dialogue*] (Oslo) 20:(2):129–134.

Westing, A.H. 1989c. Environmental security for the Danube basin. *Environmental Conservation* (Cambridge, UK) 16:323–329.

Westing, A.H. 1989d. International Baltic Sea Fishery Commission. In: Westing, A.H. (ed.). *Comprehensive Security for the Baltic: an Environmental Approach*. London: Sage Publications, 148 pp: pp 72–74.

Westing, A.H. 1989e. Our place in nature: reflections on the global carrying-capacity for humans. In: Polunin, N. & Burnett, J.H. (eds). *Maintenance of the Biosphere*. Edinburgh: Edinburgh University Press, 228 pp: pp 109–120.

Westing, A.H. 1989f. The atmosphere as a common heritage of humankind: its role in environmental security. *Environment* (Philadelphia, PA) 31(3):3–4.

Willot, A. et al. 1981. *Study on all the Aspects of Regional Disarmament*. New York: UN General Assembly, Centre for Disarmament, Disarmament Study Series No. 3 (A/35/416), 59 pp.

WRI, IIED, & UNEP. 1988–1989. *World Resources 1988–89*. New York: Basic Books, 372 pp.

Chapter 5
Regional Security: Maritime Issues

5.1 Introduction

The world ocean covers more than two-thirds of the global surface, distributed among several major and numerous minor basins. Of the present 189 nations, 150 (79 %) enjoy direct access to the ocean.[1] Although most nations with direct access to the ocean maintain at least a small navy or coast guard, less than two dozen of them maintain substantial naval forces, and only a few maintain really huge ones. However, the vast extent, enormous volume, remarkable buoyancy, low level of friction, partial opacity, and extensive resources of the ocean all combine to make it an ever more important theater of operations for both civil and military purposes.

The present Chapter analyzes environmental dimensions of maritime security. It builds upon prior work by the author (e.g., Westing 1980, 1985, 1986, 1989).

This Chapter is reproduced from the author's Entry #233 provided in Chap. 2, with the original title, 'Environmental dimensions of maritime security'. It is used here by permission of the United Nations Institute for Disarmament Research, the copyright holder, as given on 25 March 2013. The author is grateful to Jozef Goldblat for relevant discussions.

[1] **Sovereign nations**: The number of *de facto* sovereign nations in the world is for purposes of this study considered at present (mid-1992) to be 189: the 178 members of the United Nations plus Andorra, Georgia, Kiribati, Macedonia, Monaco, Nauru, Switzerland, Taiwan, Tonga, Tuvalu, and Vatican City. Of these 189 *de facto* sovereign states, 150 are coastal (littoral) (of which 46 are islands) and 39 are landlocked (hinterland). [As at February **2013**, there are perhaps 195 sovereign states (the 193 members of the United Nations plus Taiwan and Vatican City), of which 151 are coastal (including 49 which are islands) and 44 are landlocked.]

A. H. Westing, *From Environmental to Comprehensive Security*,
SpringerBriefs on Pioneers in Science and Practice 13,
DOI: 10.1007/978-3-319-00687-1_5, © The Author(s) 2013

5.2 The Ocean as a Common Natural Heritage of Humankind

Most coastal nations (ca four-fifths of them) have within the past 15 years or so laid claim to an exclusive economic zone of coastal ocean that extends out for 370 km (200 nautical miles), altogether amounting to perhaps 110 million km^2 of ocean (ca 30 % of the total ocean). Nonetheless, the ocean beyond any national jurisdiction, including the seabed beneath it, remains a vast extra-territorial domain, still covering some 250 million km^2.

It is clear to many that for a combination of social and environmental reasons, the ocean and its natural resources should be treated as a common natural heritage of humankind, that is, as an environmental domain to be equitably managed in perpetuity—managed on a sustained basis for the benefit of humankind as a whole. In fact, such a common-heritage notion was supported in principle, at least with respect to the extra-territorial seabed, by more than 100 nations as long ago as 1970 (UNGA 1970), and would gain important legal support from the 1982 Law of the Sea Convention, were that instrument to come into force.[2]

5.3 Abuses of the Ocean

Human abuses of the ocean can emanate either from the civil sector of society or from its military (including naval) sector. Most abuses emanate from the civil sector, simply because only a rather small fraction of all human activity falls within the military sector (of the order of 5 %). Moreover, most of the military activity occurs during peacetime. On the other hand, the recurring wartime activities of the military sector have a great potential for serious abuse.

So far, the continuing influx of pollutants into the ocean as a whole (much of it from land-based sources) has been rendered more or less innocuous by dilution and decomposition (both abiotic and biotic) (Goldberg 1976; McIntyre et al. 1990; Strong 1991b). On the other hand, a considerable number of local areas that are partially cut off from the rest of the ocean are being subjected to waste inputs beyond the level of sustainable discharge, for example, the Baltic Sea (Westing 1989), Mediterranean Sea (Haas 1990), and Persian Gulf (Strong 1991a). Microbial contamination from raw sewage has become a public health problem in an increasing number of coastal areas.

The harvesting of fish or other ocean species beyond the capacity of the exploited populations to renew their numbers is a flagrant abuse of the sacrosanct principle of sustained-yield management. Of the 16 recognized major marine

[2] The 1982 Law of the Sea Convention (UNTS 31363) entered into force in 1994. As at February 2013 it had 165 states parties. The present Chapter had first appeared in 1992 prior to its coming into force, thus reflecting that status.

fishery areas, 4 are now clearly being fished beyond sustainability (the Northwest Pacific, Southeast Pacific, Mediterranean plus Black, and Eastern Indian); and 5 additional ones more ambiguously so (the Northeast Pacific, East-Central Pacific, West-Central Pacific, Northeast Atlantic, and East-Central Atlantic) (WRI 1990, Table 23.3). Such over-use is to be expected in the absence of well-administered multilateral compacts owing to the well known tragedy-of-the-commons phenomenon (Hardin 1968).

Military abuses of the ocean are to a considerable extent comparable to the civil ones, but some are more or less distinct and have the potential for being more spectacular. The safe disposal of decommissioned nuclear-propelled submarines presents a particular problem. Underwater explosions and contamination with radioactive isotopes and chemical warfare agents are among the significant military abuses, both in peacetime and wartime. A recently revealed case of repeated secret radioactive dumping provides one flagrant example (Marshall 1992; Tyler 1992). Military landing operations, whether as training exercises or in wartime, have the potential for severely disrupting estuaries, coral reefs, and other important inshore habitats.

In wartime, the sinking of naval and merchant ships has the potential for releasing inherently dangerous or noxious substances, so-called dangerous forces (whether explosive, poisonous, or otherwise environmentally disruptive). Radioactive contamination from damaged nuclear propulsion systems is a most worrisome possibility, especially considering the large numbers of nuclear-powered ships in service that would become high-priority targets in wartime: currently, more than 400 submarines and more than 50 surface ships. The explosion of liquified natural gas (LNG) from damaged LNG tankers could be locally catastrophic.

Releases of oil from damaged shore facilities, pipelines, offshore platforms, and supertankers, whether with hostile intent or otherwise, are additional matters of concern. Indeed, during the Gulf War of 1991, at least 160 thousand m^3 (1 million barrels) of oil was released into the Persian Gulf, with substantial ecological impact (Sheppard and Price 1991; Strong 1991a). Regarding this, the United Nations Security Council resolved that Iraq is 'liable...for any direct...environmental damage and the depletion of natural resources' (UNSC 1991: Paragraph 16). Sea mines that become unanchored and drift out of their zone of emplacement, or those that remain functional after their military purpose has expired, are a special environmental hazard (Westing 1985: 5). Apropos the diverse military threats to the ocean environment, during the Gulf (Iran-Iraq) War of 1980–1988 the United Nations Security Council specifically called upon 'both parties to refrain from any action that may endanger...marine life in the region of the Gulf' (UNSC 1983: Paragraph 5).

Nuclear-propelled and nuclear-armed naval forces are a menace to the ocean environment even in peacetime, owing to the occasional accidents associated with training, testing, and routine patrolling (Arkin and Handler 1989; Gregory and Edwards 1989; Handler et al. 1990). To date these accidents have from time to time resulted in local radioactive contamination; and, although the risk is minimal, the potential always exists for the far greater calamity of a nuclear explosion.

The testing of nuclear weapons above, at, or beneath the ocean surface has accounted for a significant amount of radioactive contamination of the ocean (Westing 1980: 159–163), and at least the possibility exists for further abuse of this sort. The dumping of chemical warfare agents into the ocean—whether by accident or intent—has resulted in serious local contamination (Laurin 1991; Westing 1980: 163–165).

Thus it becomes clear that any measure, civil or military, that would contribute to an amelioration of the natural-resource and other environmental problems just outlined would contribute to human security, thereby providing a further justification for pursuing maritime confidence-building measures.

5.4 Confidence-Building Measures

5.4.1 Initial Considerations

A proper test of whether measures to build confidence among nations have been successful is whether the nations in question have thereby established sufficient mutual trust, and come to feel sufficiently secure, to reduce the size of their military sector. However, the ultimate test of their success must be measured in terms of the reduced frequency with which the resolution of their interstate disputes is carried out using deadly force.

In fact, at no time since World War II has the situation been more propitious, on the one hand, to conserve and restore the global biosphere; and, on the other, for the major powers to shrink their military sectors. And, in the present context, virtually every nation in the world has recently expressed a specific interest in addressing the issue of naval disarmament (UNGA 1990c). Although most attention by diplomats and scholars has in recent years been directed towards restrictions or reductions related to ground, air, and space forces, the naval sector has not been entirely ignored. Naval strengths have been scrutinized with care (e.g., Alatas et al. 1986; Arkin 1987; Durch 1991; Fieldhouse and Taoka 1989; Handler and Arkin 1990); and naval disarmament has been examined in some detail, both monographically (e.g., Alatas et al. 1986; Blechman et al. 1991; Fieldhouse 1990; Hill 1989; Lodgaard 1990; UN 1990a) and in valuable briefer treatments (e.g., Arnett 1990; Eberle 1990; Prawitz 1990; Prins 1990; Ross 1989–1990).

Measures, direct or indirect, that lead to a shrinking of the naval sector of the major powers are of potential benefit to the ocean environment inasmuch as this would reduce the routine naval peacetime disruption, and presumably also the frequency and extent of wartime disruption. First and foremost, any measure that reduces the risk of nuclear war at sea and, secondarily, of the likelihood of nuclear-weapon accidents at sea, would be a boon to the ocean environment, the importance of which is difficult to exaggerate. Thus, the recent unilateral announcement by the USA that 'under normal circumstances' its ships would no longer carry tactical nuclear weapons must be lauded (Bush 1991); as must the responses of the then USSR (Gorbachev 1991) and the United Kingdom (Schmidt 1992) to do the same. It can only be hoped that all of the major naval powers will recognize the need for, and conclude, a multilateral agreement that commits them to such repudiation. Moreover, regarding the dangers associated with naval nuclear-propulsion systems, one highly useful step would be for the International Atomic Energy Agency (Vienna) to develop suitable safety standards for all ocean-going reactors, comparable to those it has for the land-based ones, perhaps doing so in cooperation with the International Maritime Organization (London). Nations that adopt such (or even more stringent) standards would certainly be contributing to mutual trust through such action.

More generally, confidence could be enhanced by virtue of a coastal nation encouraging lines of communication among its ocean officials, both civil and military, and their counterparts in other countries; and by doing the same for its marine research institutes. Regular meetings ought to be set up for standardizing monitoring procedures and for the routine exchange of marine biological and other oceanographic information. The International Council for the Exploration of the Sea (Copenhagen) could play a valuable role in catalyzing such bureaucratic and research cooperation.

Singled out next for special discussions are confidence-building measures related to: (1) international law; (2) ecogeographical regions; (3) natural resources; and (4) nature reserves.

5.4.2 International Law

There is considerable room for building confidence in the realm of international ocean law, especially as it relates to naval disarmament (Bring 1990a; Goldblat 1990). Thus, a number of multilateral treaties are in force that serve to protect the ocean environment from one aspect or another of military disruption, but their value could be strengthened in various ways. Treaties having special potential in this regard include: (1) the 1963 Partial Test Ban Treaty (UNTS 6964), in which the states parties agree not to test nuclear weapons in the ocean environment; (2) the 1971 Seabed Treaty (UNTS 13678), in which the states parties agree not to emplace nuclear or other weapons of mass destruction in the seabed beyond 22 km (12 nautical miles) of the shoreline, an exclusion zone of about 350 million km^2 of ocean (ca 97 % of the total ocean); (3) the 1907 Sea Mine Convention, in which the states parties eschew the use of unanchored mines and additionally agree to remove any mines they have laid at the cessation of hostilities; (4) 1977 Protocol I on the Protection of Victims of International Armed Conflicts (UNTS 17512), in which the states parties agree not to employ methods of warfare which are intended, or may be expected, to cause widespread, long-term, and severe damage to the natural environment[3]; and (5) the 1967 Latin American Nuclear-weapon Free Treaty (UNTS 9068), in which the states parties have the intent to keep all nuclear weapons out of about 68 million km^2 of ocean surrounding them (ca 19 % of the total ocean).[4] Some would add to this enumeration the 1977 Environmental Modification Convention (UNTS 17119), in which the states parties agree (with some debilitating provisos) not to manipulate the ocean environment for hostile purposes (Westing 1992).

Not all eligible nations have become a state party to the five above-noted ocean-related treaties considered here to be of particular relevance (cf. Table 5.1). Thus, one way to build confidence that remains open to many nations—both coastal and landlocked—is for them to develop a recognition of the importance of these treaties sufficient to join them (concomitantly enacting any necessary coordinate domestic legislation); and, as necessary, to conclude an appropriate Safeguard Agreement with the International Atomic Energy Agency (Vienna). Those among the permanent members of the United Nations Security Council not states parties to these treaties (or their relevant protocols) could generate much confidence by joining them. Some suggestions to the contrary (e.g., Thorpe 1987), the 1907 Sea Mine Convention remains an important instrument, especially in the absence of a comprehensive new treaty of the sort (cf. Dimitrov 1992).

Three additional ocean-related treaties lend themselves well to confidence building, each in its own fashion: (1) 1959 Antarctic Treaty (UNTS 5778); (2) 1973/1978 International Convention for the Prevention of Pollution from Ships (UNTS 22484); and (3) 1982 Law of the Sea Convention (UNTS 31363).

---

[3] **1977 Protocol I** (UNTS 17512): It has been suggested that the protection of the natural environment afforded by this instrument does not apply to naval warfare (e.g., Alatas et al. 1986: 78). This is *not* the case. It is true that the protection deriving from its Article 55 is subject to the limitations imposed by Article 49; thus, although the source of the threat—whether from land, sea, or air—is not limited by Article 49, some might suggest that the protection resulting from Article 55 applies only to the terrestrial environment. On the other hand, the protection of the natural environment afforded by Article 35 has no comparable limitations attached to it, either as to origin of attack or location of target.

It is useful to point out that, by virtue of Articles 35 and 55, environmental protection has in a formal sense has become part and parcel of the International Humanitarian Law component of the Law of War or Armed Conflict. This is especially important because the environmental protection deriving from the corpus of International Environmental Law is largely inapplicable to environmental disruption of military origin (cf. Footnote 5).

[4] **Miscellaneous ocean zones:** (1) *South Pacific Ocean*: The 1985 South Pacific Nuclear-weapon Free Treaty (UNTS 24592) serves to denuclearize the territorial seas of the states parties—a coastal strip of ocean 22 km (12 nautical miles) in width for most of the states parties. The treaty additionally delineates an ocean zone surrounding the states parties which is about 118 million km^2 (ca 33 % of the total ocean), which the nuclear powers are invited to respect as a nuclear-weapon-free zone.

(2) *Indian Ocean*: The 1971 Declaration of the Indian Ocean as a Zone of Peace designates that body, about 71 million km^2 in extent (ca 20 % of the total ocean), 'for all time as a zone of peace' (UNGA 1971). Many years of negotiation on clarifying that Declaration have to date been inconclusive (e.g., UNGA 1990a, 1991a).

Table 5.1 Multilateral treaties serving to protect the ocean (selected)

Treaty	Total parties (%)	Coastal parties (%)	Landlocked parties (%)	Permanent UNSC parties (%)
1963 Partial Test Ban	63	65	56	60
1971 Seabed	49	49	46	80
1907 Sea Mine	20	21	13	60
1977 Protocol I	59	62	49	40
1967 Latin American	70	68	100	(100)
1959 Antarctic	23	26	11	100
1973/1978 Marpol	40	47	13	100
1982 Law of the Sea	27	31	13	0

Notes

1. This Table is based on the current 189 sovereign states, of which 150 are coastal (littoral) and 39 are landlocked (hinterland) (cf. Footnote 1). The permanent members of the UN Security Council (UNSC) are China, France, Russia, United Kingdom, and USA
2. The eight multilateral treaties presented are: (1) 1963 Partial Test Ban Treaty (UNTS 6964); (2) 1971 Seabed Treaty (UNTS 13678); (3) 1907 Hague Convention VIII relative to the Laying of Automatic Submarine Contact Mines; (4) 1977 Protocol I on the Protection of Victims of International Armed Conflicts (UNTS 17512); (5) 1967 Latin American Nuclear-weapon Free Treaty (UNTS 9068), open to 33 states (31 coastal, 2 landlocked), and which the 5 permanent members of the UNSC are invited to respect, and with an International Atomic Energy Agency (Vienna) Safeguard Agreement mandated; (6) 1959 Antarctic Treaty (UNTS 5778), with 179 states automatically eligible (142 coastal and 37 landlocked) and the 10 other states eligible only by invitation of the states parties; (7) 1977 International Convention for the Prevention of Pollution from Ships ('Marpol') (UNTS 22484); and (8) 1982 Law of the Sea Convention (UNTS 31363)
3. Information on the states parties to the treaties is from the respective depositaries as of mid-1992. The texts of the major disarmament treaties can be viewed elsewhere (e.g., Goldblat 1982), as can summaries of the major environmental treaties (e.g., UNEP 1991), and a catalog of the ocean protection treaties (e.g., UN 1990b)

The 1959 Antarctic Treaty is of potential importance in the present context because its zone of application includes about 21 million km^2 of ocean (ca 7 % of the total ocean). Although military activities are prohibited in the terrestrial portion of the treaty's zone of application (cf. its Article 1), it seems clear that such a stricture does not apply to the ocean portion (cf. its Article 6). Indeed, naval operations could occur in these waters (Morgan 1990). Thus, one way in which the full ('consultative') states parties to the treaty could build confidence would be to amend the treaty by deleting Article 6. Short of such a measure, any individual present or future state party, whether full or partial ('non-consultative'), could build confidence, as appropriate, by attaching a reservation to its ratification or accession which renounces that article, or else by making a unilateral declaration to the same effect.

The 1973/1978 International Convention for the Prevention of Pollution from Ships (UNTS 22484) (for which the International Maritime Organization [London] serves as the secretariat) is one of a number of generally laudable multilateral treaties serving to minimize ocean pollution. The rather weak support it has to date received from the full range of the international community is therefore to be regretted (cf. Table 5.1). In the present context, however, attention must be drawn to the fact that this treaty does not apply to warships or any other naval ships (cf. its Article 3.3). Again, two possible ways the states parties to this treaty could build confidence would be to amend the treaty by deleting Article 3.3 (or at least by making it applicable only during wartime, and then perhaps only in a battle zone); or, short of such a measure, any individual present or future state party could build confidence, as appropriate, by attaching a reservation to its ratification or accession which renounces that article, or else by making a unilateral declaration to the same effect.[5]

The 1982 Law of the Sea Convention (UNTS 31363) provides a comprehensive and enforceable body of international law for the conservation and sustainable utilization of the ocean, both the exclusive economic zones and the extra-territorial domains beyond (the 'Area'). Included are procedures for the non-violent resolution of disputes. The environmental merits of this treaty have been widely endorsed (e.g., Brundtland et al. 1987: 272–274; IUCN et al. 1991: 160). It is thus a pity that insufficient numbers of nations—especially insufficient numbers of industrialized nations—feel comfortable with a common-heritage approach to the exploitation of natural resources in the common domains of the world (cf. Table 5.1). Therefore, as weak or peripheral as the disarmament aspects of the treaty might be (e.g., Boczek 1989; Bring 1990b; Pinto 1992), an enormous amount of confidence would be generated among nations if a sufficient number of industrialized and other nations—including landlocked nations—were to ratify this treaty. (The fact that some non-ratifying states selectively respect portions of this treaty is a mixed blessing, because such picking and choosing serves to undermine the rule of law.)

Nations could greatly enhance the level of confidence they project beyond their borders through their participation in ocean-protection treaties and by means of domestic educational programs. To borrow the exhortation from 1977 Protocol I, all states parties to these treaties should '…undertake, in time of peace as in time of armed conflict, to disseminate the [ocean-related treaties in question] as widely as possible in their respective countries and, in particular, to include the study

[5] **Sovereign immunity**: The 1973/1978 International Convention for the Prevention of Pollution from Ships (UNTS 22484) does not apply to warships or any other naval ships (cf. its Article 3.3). Regrettably, such so-called sovereign immunity is a common feature of ocean-related treaties. For example, the 1972 Convention on the Prevention of Marine Pollution by Dumping of Wastes and other Matter (UNTS 15749) has a similar restriction (cf. its Article 7.4). And the 1982 Law of the Sea Convention (UNTS 31363) repeats such a restriction in a number of its articles. Of special importance is the deleterious waiver spelled out in its Article 236: 'The provisions of this Convention regarding protection and preservation of the marine environment do not apply to any warship, naval auxiliary, other vessels or aircraft owned or operated by a State and used, for the time being, only on government non-commercial service'. A detailed exposition of sovereign immunity is available (Pinto 1992).

thereof in their programmes of military instruction and to encourage the study thereof by the civilian population…' (Article 83.1). The International Committee of the Red Cross (Geneva) could assist in this process of strengthening the cultural norms that underpin the legal norms, by developing appropriate curricula and study guides, as could the United Nations Educational, Scientific and Cultural Organization (Paris).

5.4.3 Ecogeographical Regions

Most natural-resource and other environmental problems are essentially limited to their ecogeographical region or ecosystem, a condition that holds true both for the land and the ocean. At the same time, a majority of the recurring international conflicts is between neighboring nations. It thus becomes eminently sensible to foster cooperation among a group of nations that shares such an ecogeographical region (often a 'sub-region' in political terminology).

With specific reference to the ocean, the best candidates for regional confidence building are the nations that together occupy the drainage basin (catchment basin, watershed) of a semi-enclosed sea. Indeed, efforts along these lines are being actively pursued in a number of such regions, for example, the Mediterranean Sea (Haas 1990), the Baltic Sea (Westing 1989), and the North Sea (Sætvik 1988). The United Nations Environment Programme (Nairobi) has been singularly successful with its 'Regional Seas' program, in which it has catalyzed confidence-building measures among the nations associated with at least eight different semi-enclosed portions of the ocean—in some instances, despite a history of local political antagonisms (Gebremedhin 1989). Additional regions are ripe for such attention, among them the Arctic Ocean and the South China Sea.

5.4.4 Natural Resources

Ocean fisheries are becoming ever more important in concert with rising human numbers and aspirations. The fish and other species that comprise this major renewable natural resource are harvested primarily over the continental shelves and thus largely (ca 90 %) within the exclusive economic zones that have been proclaimed in recent decades. Although most national fishing fleets are coastal, a dozen or more nations maintain distant-water fleets, several of which are very large. As noted earlier, various of the major marine fishery areas (large marine ecosystems) are currently being abused, that is, utilized beyond their sustainable yield. The nations exploiting such an area—not to mention the biota being exploited—would benefit from scientifically (ecologically) based cooperative efforts, both informal and formal, with sustained-yield management as their aim (Belsky 1990; Peterson and Teal, 1986; Young 1989).

Conflicts arise when foreign fishing vessels encroach upon a nation's exclusive economic zone, a common occurrence. Fishing on the high seas (i.e., in extra-territorial waters) can also lead to conflict, for example, when a nation is seen to be harvesting beyond the level of renewal. There is also a potential for conflict when particular fish stocks overlap the border between the exclusive economic zones of two countries, or the border between an exclusive economic zone and the high seas—situations that, in fact, often prevail. Littoral states patrol their exclusive economic zones with their coast guards, with special fishery police, with their armed forces (both naval and air), and with combinations thereof. However, many developing countries are incapable of carrying out fully adequate policing and might thus consider instituting cooperative (and thus confidence-building) policing arrangements with their neighbors.

Considerable numbers of multilateral treaties exist (often with associated fishery commissions) that have the purpose of managing either a particular category of fish (e.g., the 1949 Inter-American Tropical Tuna Commission) or a particular portion of the ocean (e.g., the 1952 International North Pacific Fisheries Commission). Fishing nations thus have the opportunity to maintain the sustained yield of such fish or regions via these commissions. However, all relevant nations would have to become party to the treaties. Moreover, the states parties to such a treaty would have to provide the commission with adequate research capabilities to establish proper fishing limits. Even more important, they would have to empower the commission to allocate fishing quotas among themselves, and to monitor and police the harvesting. Most of these commissions are in practice quite ineffectual so that confidence would be generated among the involved nations to the extent that they truly cooperated in these worthy endeavors. Finally, the 1982 Law of the Sea Convention, once widely adopted, could become the single most important multilateral treaty related to ocean fisheries: it would provide for the sustainable utilization of the renewable natural resources of the entire ocean, both within and beyond the zones of national jurisdiction, thus in essence treating those resources as a common heritage of humankind.

The special case of semi-enclosed portions of the ocean, which provides great opportunity for confidence building among the nations that share such an eco-geographical region, has already been examined above.

Numerous bilateral treaties also exist, often between neighboring littoral states, for the purpose of avoiding fishing conflicts, thereby fostering trust between them. One interesting example is the 1975/1976 Exchange between Kenya and Tanzania on the Territorial Sea Boundary (UNTS 15603) that has established a common fishing zone extending out for 22 km (12 nautical miles) on either side of their sea boundary.

Finally, it is worth noting that particularly egregious assaults on the renewable natural resources of the ocean beyond the limits of national jurisdiction can lead to widespread outrage and coordinated international condemnation. The best recent example such confidence-building action has been the reasonably successful attempt by the community of nations to put an end to drift-net fishing on the high seas (UNGA 1989, 1990b, 1991b).

5.4.5 Nature Reserves

An even less adequate worldwide fraction of the marine environment than of the terrestrial environment has been set aside as nature reserves: less than 0.2 % of the ocean (with a substantial portion of the protected ocean accounted for by one site, the Australian Great Barrier Reef Marine Park) as opposed to about 4 % of the land.[6] Given the ever greater pressures on ocean biota and ocean ecosystems noted earlier, it becomes increasingly urgent to set aside additional ocean refugia (IUCN et al. 1991: 157–158; Lien and Graham 1985; Salm and Clark 1984; Tisdell and Broadus 1989). Comparable to the case of terrestrial reserves, representative coastal and other marine ecosystems must be safeguarded as sources of assured replenishment for the ocean's renewable natural resources (some of which have become highly endangered) and, more generally, to maintain the genetic diversity of the ocean biota. To borrow a relevant guideline from the 1982 Law of the Sea Convention, 'The measures…shall include those necessary to protect and preserve rare or fragile ecosystems as well as the habitat of depleted, threatened or endangered species and other forms of marine life' (Article 194.5). Additional justifications for establishing marine (and other) nature reserves include their value for comparative purposes in ecological research, their recreational and aesthetic values, and in some instances their archeological or other cultural values.

No matter where in the ocean such nature reserves are established, they would be of widespread benefit. It would thus be in the enlightened self-interest of the wealthier nations of the world not only to establish new reserves within their own territorial waters and exclusive economic zones, but additionally to assist the poorer nations to do likewise. Such cooperation could take the form of scientific, technological, and financial support; and it could be carried out bilaterally or else via such agencies as the United Nations Environment Programme (Nairobi) or the International Union for Conservation of Nature (Gland, Switzerland). For any nature reserve created in the extra-territorial domain of the ocean to succeed, a widely supported multilateral treaty plus administration by an intergovernmental agency would seem to be necessary.

There is no question that the value to nature of a global system of protected areas in the ocean on the one hand, and the enhancement of international confidence engendered by such a system on the other, would outweigh the modest concomitant curtailment in freedom of the seas.

[6] Formally protected areas worldwide as of September 2012: 1.6 % of the ocean (although ca 7 % of territorial waters), and 12.7 % of the land (cf. http://www.unep-wcmc.org/ppr2012_903.html).

5.5 Recommendations

Maritime confidence-building measures associated with the environment can take a number of forms; and they can have various outcomes. To begin with, it has been suggested here that confidence among nations could be greatly enhanced by making the ocean nuclear-weapon free, the latter the potentially single greatest boon to the ocean environment. Measures taken to avoid radioactive releases from the many naval nuclear propulsion systems, during both peacetime and wartime, would generate additional confidence.

Confidence could be enhanced to the extent that more nations—both coastal and landlocked—recognize the need to become party to the various major multilateral treaties that contribute to the protection of the ocean environment from military activities; and then, after joining these treaties, go on to uphold and support them through co-ordinate domestic legislation, by educational means, and in other ways. Confidence could be further enhanced if the states parties to certain multilateral ocean-related treaties agreed to expand the scope of those treaties to encompass military activities; or, short of such improvement, for individual nations to make unilateral commitments to that effect.

An enormous level of confidence would be generated if sufficient numbers of industrialized and other nations were to embrace the 1982 Law of the Sea Convention. They would thereby be supporting the rule of law, but more specifically, they would be accepting the concept that the ocean is subject to a comprehensive body of law. They would also be accepting that at least some of its components (the extra-territorial seabed and, to some extent, all fish) are a common natural heritage of humankind—as well as the practical implications, both societal and environmental, of such acceptance.

Many advantages would accrue to the nations that share the drainage basin of a semi-enclosed portion of the ocean if they were to cooperate on its environmental protection and sustainable exploitation through impartial mechanisms of monitoring and enforcement; and the level of confidence among them would be strengthened in the process.

The establishment of a co-ordinated system of nature reserves in the ocean—including, as necessary, the transfer of technology and other cooperation between rich and poor nations—would be of great value to the ocean ecology and would create much confidence among the cooperating nations.

5.6 Conclusion

Ocean ecosystems, especially in coastal and semi-enclosed regions, are coming under increasing human pressure (McIntyre et al. 1990; Strong 1991b). Some of this pressure emanates on a continuing basis from the military sector of society; and the possibility always exists for hostile actions that lead to locally serious or even cataclysmic disruption (Alatas et al. 1986). In addition, there is much, and growing, room for international conflict over ocean resources (Peterson and Teal 1986). Thus it becomes ever more urgent for the community of nations to develop sufficient mutual trust to carry out a process of naval and other disarmament, to facilitate cooperation in the equitable sharing of the ocean's natural resources in perpetuity—doing so for the benefit of humankind and the biosphere upon which it depends.[7]

References

Alatas, A. et al. 1986. *The Naval Arms Race* New York: UN Department for Disarmament Affairs, Disarmament Study Series No. 16 (A/40/535): 100 pp.

Arkin, W.M. 1987. *The Nuclear Arms Race at Sea*. Washington: Greenpeace & Institute for Policy Studies, Neptune Paper No. 1, 46 pp.

Arkin, W.M. & Handler, J. 1989. *Naval Accidents 1945–1988*. Washington: Greenpeace and Institute for Policy Studies, Neptune Paper No. 3, 84 pp. [Updated by: Handler et al., 1990, cf. below.]

Arnett, E.H. 1990. *Naval Nuclear Weapons*. Washington: American Association for the Advancement of Science, Publication No. 90–51S, 12 pp.

Belsky, M.H. 1990. Interrelationships of law in the management of large marine ecosystems. In: Sherman, K. et al., (eds): *Large Marine Ecosystems: Patterns, Processes, and Yields*. Washington: American Association for the Advancement of Science, 242 pp: pp 224–233.

Blechman, B.M. et al. 1991. *Naval Arms Control: A Strategic Assessment*. New York: St Martin's Press, 268 pp.

Boczek, B.A. 1989. Peaceful Purposes Provisions of the United Nations Convention on the Law of the Sea. *Ocean Development & International Law* (New York) 20:359–389. [Similar in: *Ocean Yearbook* (Chicago) 8:329–361. 1990.]

Bring, O. 1990a. International law and arms restraint at sea. In: Lodgaard, S. (ed.). *Naval Arms Control*. London: Sage Publications, 264 pp: pp 187–197.

Bring, O. 1990b. Naval Arms Control and the Convention on the Law of the Sea. In: Fieldhouse, R. (ed.). *Security at Sea: Naval Forces and Arms Control*. Oxford: Oxford University Press, 301 pp: pp 137–143.

Brundtland, E.H. et al. 1987. *Our Common Future: World Commission on Environment and Development*. rev. edn. Oxford: Oxford University Press, 400 pp.

Bush, G. 1991. *New initiatives to reduce US nuclear forces. US Department of State Dispatch* (Washington) 2:715–718. [Also in: *Arms Control Today* (Washington) 21(8):3–5. 1991.]

Dimitrov, G. 1992. Possible new restrictions on the use of naval mines. In: Goldblat, J. (ed.). *Maritime Security: the Building of Confidence*. UN Institute for Disarmament Research, Document No. UNIDIR/92/89, 159 pp. pp 79–90.

Durch, W.J. 1991. Naval arms control: tons of options? *Arms Control Today* (Washington) 21(9):7–12.

Eberle, J. 1990. Global security and naval arms control. *Survival* (London) 32:325–332.

[7] It can only be hoped that the independent *Global Ocean Commission* (Oxford) established in February 2013, will facilitate success in addressing some of these pressing issues inasmuch as its mission is to formulate politically and technically feasible recommendations to address: (1) overfishing; (2) large-scale loss of habitat and biodiversity; (3) the lack of effective management and enforcement; and (4) deficiencies in high-seas governance (cf. www.globaloceancommission.org).

Fieldhouse, R. (ed.). 1990. *Security at Sea: Naval Forces and Arms Control* Oxford: Oxford University Press, 301 pp.
Fieldhouse, R. & Taoka, S. 1989. *Superpowers at Sea: an Assessment of the Naval Arms Race.* Oxford: Oxford University Press, 183 pp.
Gebremedhin, N. 1989. Lessons from the UNEP Regional Seas Programme. In: Westing, A.H. (ed.). *Comprehensive Security for the Baltic: an Environmental Approach.* London: Sage Publications, 148 pp: pp 90–98.
Goldberg, E.D. 1976. *Health of the Oceans.* Paris: UNESCO Press, 172 pp.
Goldblat, J. 1982. *Agreements for Arms Control: a Critical Survey.* London: Taylor & Francis, 387 pp.
Goldblat, J. 1990. Seabed Treaty and arms control. In: Fieldhouse, R. (ed.). *Security at Sea: Naval Forces and Arms Control.* Oxford: Oxford University Press, 301 pp: pp 187–200.
Gorbachev, M.S. 1991. [Speech of 5 October 1991]. *Current Digest of the Soviet Press* (Columbus, OH) 43(40):13–14. [Also in: *Arms Control Today* (Washington) 21(8):6. 1991.]
Gregory, S. & Edwards, A. 1989. Hidden cost of deterrence: nuclear weapons accidents 1950–88. *Bulletin of Peace Proposals* [now *Security Dialogue*] (Oslo) 20:3–26.
Haas, P.M. 1990. *Saving the Mediterranean: the Politics of International Environmental Co-operation.* New York: Columbia University Press, 303 pp.
Handler, J. & Arkin, W.M. 1990. *Nuclear Warships and Naval Nuclear Weapons 1990: a Complete Inventory.* rev. edn. Washington: Greenpeace, Neptune Paper No. 5, 108 pp. [Replaces: Handler & Arkin (1988) Neptune Paper No. 2.]
Handler, J. et al. 1990. *Naval Safety 1989: the Year of the Accident.* Washington: Greenpeace, Neptune Paper No. 4, 26 pp. [Update of: Arkin & Handler, 1989. Neptune Paper No. 3.]
Hardin, G. 1968. Tragedy of the commons. *Science* (Washington) 162:1243–1248.
Hill, J.R. 1989. *Arms Control at Sea.* Annapolis, MD: Naval Institute Press, 229 pp.
IUCN, UNEP, & WWF. 1991. *Caring for the Earth: a Strategy for Sustainable Living.* Gland, Switzerland: International Union for Conservation of Nature, 228 pp.
Laurin, F. 1991. Scandinavia's underwater time bomb. *Bulletin of the Atomic Scientists* (Chicago) 47(2):11–15.
Lien, J. & Graham, R. (eds). 1985. *Marine Parks and Conservation: Challenge and Promise.* Toronto: National & Provincial Parks Association of Canada [now Canadian Parks & Wilderness Society], 2 vols (254+211 pp).
Lodgaard, S. (ed.). 1990. *Naval Arms Control.* London: Sage Publications, 264 pp.
Marshall, E. 1992. Scramble for data on Arctic radioactive dumping. *Science* (Washington) 257:608–609.
McIntyre, A.D. et al. 1990. *State of the Marine Environment.* Nairobi: UN Environment Programme, Regional Seas Reports & Studies No. 115 (GESAMP No. 39), 111 pp.
Morgan, J.R. 1990. Naval operations in the Antarctic region: a possibility? *Ocean Yearbook* (Chicago) 8:362–377.
Peterson, S.B. & Teal, J.M. 1986. Ocean fisheries as a factor in strategic policy and action. In: Westing, A.H. (ed.). *Global Resources and International Conflict: Environmental Factors in Strategic Policy and Action.* Oxford: Oxford University Press, 280 pp: pp 114–142.
Pinto, M.C.W. 1992. The 1982 UN Convention on the Law of the Sea. In: Goldblat, J. (ed.). *Maritime Security: the Building of Confidence.* UN Institute for Disarmament Research, Document No. UNIDIR/92/89, 159 pp: pp 9–53.
Prawitz, J. 1990. Application of confidence-building measures to a naval nuclear environment. In: UN (ed.). *Naval Confidence-building Measures.* New York: UN Department for Disarmament Affairs, Disarmament Topical Paper No. 4, 320 pp: pp 117–123.
Prins, G. 1990. United Nations and naval power in the post-cold-war world. In: UN (ed.). *Naval Confidence-building Measures.* New York: UN Department for Disarmament Affairs, Disarmament Topical Paper No. 4, 320 pp 219–251. [Similar in: *Disarmament* (New York) 14(1):183–215, 1991. Also similar in: Bulletin of Peace Proposals [now Security Dialogue] (Oslo) 22:135–155, 1991.]

Ross, M.L. 1989–1990. Disarmament at sea. *Foreign Policy* (Washington) 1989-1990(77):94–112.

Sætvik, S. 1988. *Environmental Co-operation between the North Sea States: Success or Failure?* London: Belhaven Press, 173 pp.

Salm, R.V. & Clark, J.R. 1984. *Marine and Coastal Protected Areas: a Guide for Planners and Managers*. Gland, Switzerland: International Union for Conservation of Nature, 302 pp.

Schmidt, W.E. 1992. British are planning to remove A-arms from ships and aircraft. *New York Times* 141(48,599):A14. 16 June 1992.

Sheppard, C. & Price, A. 1991. Will marine life survive the Gulf War? *New Scientist* (London) 129(1759):36–40.

Strong, M.F. 1991a. *Environmental Assessment of the Gulf Crisis*. New York: UN General Assembly, Document No. A/C0NF.151/PC/72, 16 July 1991, 13 pp. [Includes reports by the International Maritime Organization (London) & the UN Environment Programme (Nairobi), the latter also separately published as Document No. UNEP/GC.16/4/Add.1, 10 May 1991,10 pp.]

Strong, M.F. 1991b. *Protection of Oceans, all Kinds of Seas including Enclosed and Semi-enclosed Seas, Coastal Areas and the Protection, Rational Use and Development of their Living Resources: Background Document*. New York: UN General Assembly, Document No. A/CONF.151/PC/69, 4 July 1991, 76 pp.

Thorpe, A.G.Y. 1987. Mine warfare at sea: some legal aspects of the future. *Ocean Development & International Law* (New York) 18:255–278.

Tisdell, C. & Broadus, J.M. 1989. Policy issues related to the establishment and management of marine reserves. *Coastal Management* (London) 17(1):37–53.

Tyler, P.E. 1992. Soviets secret nuclear dumping causes worry for Arctic waters. *New York Times* 141(48, 596):A1, A8. 4 May 1992.

UN (ed.). 1990a. *Naval Confidence-building Measures*. New York: UN Department for Disarmament Affairs, Disarmament Topical Paper No. 4, 320 pp.

UN (ed.). 1990b. *The Law of the Sea: Protection and Preservation of the Marine Environment: Repertory of International Agreements*. New York: UN Office for Ocean Affairs & the Law of the Sea, 95 pp.

UNEP (ed.). 1991. *Register of International Treaties and other Agreements in the Field of the Environment*. Nairobi: UN Environment Programme, Report No. UNEP/GC.16/Inf.4, 248 pp.

UNGA, 1970. *Declaration of Principles Governing the Sea-bed and the Ocean Floor, and the Subsoil thereof, beyond the Limits of National Jurisdiction*. New York. UN General Assembly, Resolution No. 2749(XXV), 17 December 1970, 2 pp. 108 (85 %) in favor, 14 abstentions, 0 against, 5 absent = 127]

UNGA. 1971. *Declaration of the Indian Ocean as a Zone of Peace*. New York: UN General Assembly, Resolution No. 2832(XXVI), 16 December 1971, 1 p. [60 (45 %) in favor, 56 abstentions, 0 against, 16 absent = 132]

UNGA. 1989. *Large-scale Pelagic Driftnet Fishing and its Impacts on the Living Marine Resources of the World's Oceans and Seas*. New York: UN General Assembly, Resolution No. 44/225, 22 December 1989, 3 pp. [159; adopted without a vote]

UNGA. 1990a. *Implementation of the Declaration of the Indian Ocean as a Zone of Peace*. New York: UN General Assembly, Resolution No. 45/77, 12 December 1990, 3 pp. [128 (81%) for, 17 abstentions, 4 against, 10 absent = 159]

UNGA. 1990b. *Large-scale Pelagic Driftnet Fishing and its Impact on the Living Marine Resources of the World's Oceans and Seas*. New York: UN General Assembly, Resolution No. 45/197, 21 December 1990, 3 pp. [159; adopted without a vote]

UNGA. 1990c. *Naval Armaments and Disarmament*. New York: UN General Assembly, Decision No. 45/416, 4 December 1990, 1 p. [152 (96%) in favor, 0 abstentions, 1 against, 6 absent = 159]

UNGA. 1991a. *Implementation of the Declaration of the Indian Ocean as a Zone of Peace*. New York: UN General Assembly, Resolution No. 46/49, 9 December 1991, 3 pp. [166; adopted without a vote]

UNGA. 1991b. *Large-scale Pelagic Drift-net Fishing and its Impact on the Living Marine Resources of the World's Oceans and Seas*. New York, UN General Assembly, Resolution No. 46/215, 20 December 1991, 3 pp. [166; adopted without a vote]

UNSC. 1983. *The Situation between Iran and Iraq*. New York: UN Security Council, Resolution No. 540(1983), 31 October 1983, 1 p. [12 (80 %) in favor, 3 abstentions, 0 against, 0 absent = 15]

UNSC. 1991. *[Resolution relating to the Restoration to Kuwait of its Sovereignty.]* New York: UN Security Council, Resolution No. 687(1991), 3 April 1991, 10 pp. [12 (80 %) in favor, 2 abstentions, 1 against, 0 absent = 15]

Westing, A.H. 1980. The Ocean. In: Westing, A.H. *Warfare in a Fragile World: Military Impact on the Human Environment*. London: Taylor & Francis, 249 pp: pp 144–182. [Similar in: Ocean Yearbook (Chicago) 1:436–466, 1978.]

Westing, A.H. (ed.). 1985. *Explosive Remnants of War: Mitigating the Environmental Effects*. London: Taylor & Francis, 141 pp.

Westing, A.H. (ed.). 1986. *Global Resources and International Conflict: Environmental Factors in Strategic Policy and Action*. Oxford: Oxford University Press, 280 pp.

Westing, A.H. (ed.). 1989. *Comprehensive Security for the Baltic: an Environmental Approach*. London: Sage Publications, 148 pp.

Westing, A.H. 1992. Environmental Modification Convention: 1977 to the present. In: Burns, R.D. (ed.). *Encyclopedia of Arms Control and Disarmament*. New York: Charles Scribner's Sons, 1692 pp: pp 947–954.

WRI. 1990. *World Resources 1990–91: a Report by the World Resources Institute*. 4th edn. New York: Oxford University Press, 383 pp.

Young, O.R. 1989. Comparative statics: regimes for the marine fisheries and deep-seabed mining. In: Young, O.R. *International Co-operation: Building Regimes for Natural Resources and the Environment*. Ithaca: New York: Cornell University Press, 248 pp: pp 109–144.

Chapter 6
Regional Security: Transfrontier Cooperation

6.1 Introduction

To achieve comprehensive human security requires the satisfaction of both of its two somewhat intertwined components, social security (with its political, military, economic, and ethical elements) and environmental security (with its utilization-oriented and protection-oriented elements) (Westing 1989a). Unfortunately, the quest for comprehensive human security is becoming ever more elusive. Important among the grave threats to this aspiration are: (a) the continued resort to deadly violence for the resolution of interstate disputes; and (b) the rapid loss of the genetic resources (biodiversity) of the world. The focus of the present analysis is on the feasibility and efficacy of ameliorating both of these two disparate threats to human security simultaneously, doing so through the establishment and maintenance of jointly administered cross-border protected natural areas.

In other words, the subject being addressed here is the extent to which a jointly administered, demilitarized, nature-protected buffer zone straddling an interstate border might not merely help to conserve biodiversity, but at the same time serve as a political confidence-building measure (*sensu* Pfeiffer et al. 1982) between contiguous nations. To that end, the present Chapter examines the current situation regarding border and other interstate wars on the one hand, and regarding the genetic resources (biodiversity) of the world on the other. Some guidelines are provided for the establishment and maintenance of transfrontier reserves. Looked at also are existing precedents for demilitarized border regions and for transfrontier protected natural areas. Finally, a number of disparate sites in several parts of the

---

This Chapter is reproduced from the author's Entry #245 provided in Chap. 2, with the original title, 'Building confidence with transfrontier reserves: the global potential'. It is used here by permission of the United Nations Environment Programme, the copyright holder, as given on 18 March 2013. The author is grateful to Naigzy Gebremedhin and Carol E. Westing for their guidance. A preliminary version of this Chapter was presented at 'The IVth World Congress on National Parks and Protected Areas', Caracas, 10–21 February 1992, of the International Union for Conservation of Nature (Gland, Switzerland). All International Union for Conservation of Nature species threat categories have been updated to March 2013 (cf. www.iucnredlist.org).

A. H. Westing, *From Environmental to Comprehensive Security*,
SpringerBriefs on Pioneers in Science and Practice 13,
DOI: 10.1007/978-3-319-00687-1_6, © The Author(s) 2013

world are suggested as likely locations for transfrontier reserves that would serve the dual purposes suggested above. A bibliography of relevant publications is provided elsewhere as well (Westing 1993d).

6.2 Interstate Wars

Turning first to the border and other interstate wars in the world, it is a sad fact that these continue to occur with no readily apparent change in frequency (Eckhardt 1991; Westing 1982). The land areas of the world (with the possible exception of Antarctica) are currently divided into about 191 tenaciously sovereign nations, most of them sharing one to several borders with each other.[1] Interactions between these many contiguous nations vary with time from amicable to hostile, with the hostile interactions all too often escalating to deadly violence. Scores of interstate wars have been waged just since World War II, mostly in the Third World. These many wars have resulted in human death and suffering on a grand scale, in the squandering of scarce material and financial resources, and in substantial damage to the environment (Westing 1980, 1984, 1988). Protected natural areas are often among the sites disrupted by military actions (Westing 1992).

Enmity between contiguous nations can have a complex and often obscure history. Indeed, a seemingly infinite series of scholarly books and articles has grappled with the causes of war. Even the many seemingly circumscribed so-called border wars can be the result of interstate animosities that have little to do with the shared border itself (Grundy–Warr 1990; Kratochwil 1986–1987; Kratochwil et al. 1985; Mandel 1980; Prescott 1987; Tägil 1982; Tägil et al. 1977). Nonetheless, it is important to point out that more than half of the nations of the world share borders that are ill-defined, or even contested, in whole or in part. The resulting border disputes often remain a continuing source of friction between contiguous nations for years, and become at least the proximate cause of many of the border wars (Day 1987; Downing 1980; Starr and Most 1983). Mechanisms do, of course, exist for the peaceful resolution of border disputes, via the good offices of the International Court of Justice (The Hague), the Permanent Court of Arbitration (The Hague), and other tribunals; but it is clear that these are not sufficiently often employed (Cukwurah 1967; Sharma 1976).

The territorial extents of the nations of the world—and thus the locations of their borders—are, of course, the result of a potpourri of geographical determinants and past human actions, both non-violent and violent. Thus, the national divisions of today may coincide to some greater or lesser extent with ethnic, religious, linguistic, or ideological groupings. What they do not very often

[1] As at February 2013, there are perhaps 195 sovereign states (the 193 members of the United Nations plus Taiwan and Vatican City).

coincide with is units determined by ecological factors, that is, to ecosystems or ecogeographical regions.

The significance of the just-noted relatively frequent lack of congruity between political units (i.e., nations) and ecological units (i.e., ecogeographical regions) is profound: the utilization of many of the global natural resources and the abatement of much of the global air and water pollution must, for rational management, become the joint endeavor of the disparate nations that happen to share any particular ecogeographical region (Vallentyne and Beeton 1988; Westing 1989b, c). Although such natural-resource utilization (whether for sustained yield or for sustained discard) is a crucial component of comprehensive regional security (Westing 1989a), the point to be made here is that the protection of genetic resources often requires similarly shared action across national borders. Indeed, without joint management a shared (transboundary) ecosystem might well lead to strife. In either case, the need for joint interstate action in environment-related issues presents a challenge to the widespread strongly held notions of national sovereignty (Dupuy 1982; Peters et al. 1989; Schrijver 1988). Thus, the establishment and administration of any transfrontier protected natural area must be sensitive to this issue.

6.3 Genetic Resources (Biodiversity)

Turning now to the dwindling genetic resources of the world, it is another sad fact that ever greater numbers of species of plants and animals are being driven to extinction (Ehrlich and Ehrlich 1981; McNeely et al. 1990; Myers 1979; Reid and Miller 1989; Wolf 1987). These extirpations result largely from the ever intensifying competition for the global land and its resources between humankind on the one hand and all other living things on the other. As human numbers and needs multiply, the remaining flora and fauna inexorably decrease apace. To help save the remaining genetic resources of the world—so necessary for the well-being both of the biota and of humans—will require a great expansion of the global land area devoted to protected natural areas, an expansion to perhaps three times the current level (Westing 1990a). Today a mere 4 % is formally protected (IUCN 1990b).[2] Indeed, only about 23 nations formally reserve an adequate 10 % or more for nature, whereas about 99 nations reserve a trivial 1 % or less (WRI 1992, Table 20.1). Thus, one important goal must be to elucidate to the satisfaction of decision makers the necessity for expanded efforts to conserve genetic resources as an obligate component of environmentally sound and sustainable development (cf., e.g., IUCN et al. 1991).

[2] Formally protected areas worldwide as of September 2012: 1.6 % of the ocean (although ca 7 % of territorial waters), and 12.7 % of the land (cf. http://www.unep-wcmc.org/ppr2012_903.html).

The possibilities for contiguous nations to enter into a joint venture to establish a demilitarized border-straddling protected natural area are, of course, rather limited. To begin with, regional disarmament is not a widely accepted or readily achieved status (Willot et al. 1981). And then, of course, numerous of the ill-defined or contested border regions (or contested oceanic islands) are urbanized, devoted to agriculture, industrialized, militarized, or otherwise unsuitable for consideration as a protected natural area. Indeed, some of the contested border regions (or contested oceanic islands) might be considered by the disputing nations to involve their 'supreme' interests, for example, if they contained vital resources such as oil or water, if they were inhabited by ethnically or linguistically related people, or if they had militarily strategic importance. On the other hand, some border regions (and oceanic islands) are without traditionally valuable resources and are at the same time still relatively wild, uninhabited (or almost so), and thus 'remote'. When such areas are contested—or ostensibly contested—the dispute may be motivated largely by national pride or perhaps by unrelated antagonisms between the states.

It becomes clear from the foregoing that it is important: (a) to establish criteria for jointly administered transfrontier protected natural areas; and (b) to survey the border regions of the world with the aim of identifying those that might, in fact, be suitable candidates for such administration. The suggested survey should, of course, not be limited to presently unfriendly contiguous nations. The joint establishment of protected natural areas by friendly nations would seem to serve at least three valuable purposes: (a) it would contribute to the conservation of genetic resources; (b) it would probably serve as an example to other nations of the feasibility of such joint ventures; and (c) it might contribute to cementing the existing amicable relationship between the contiguous nations.

6.4 Establishment and Maintenance

To establish a transfrontier reserve, it is necessary for the potential partner countries to begin informal discussions about the desirability and modalities of such collaboration, covering both the political and environmental dimensions. From the standpoint of nature protection, at some point those discussions would benefit from the advice of such agencies as the United Nations Environment Programme (Nairobi) or the International Union for Conservation of Nature (Gland, Switzerland). Similarly, from the standpoint of confidence building (including reserve demilitarization), at some point those discussions would benefit from the advice of such agencies as the United Nations Institute for Disarmament Research (Geneva) and the International Committee of the Red Cross (Geneva). There are, of course, further agencies that could be useful to draw upon in support of these discussions, among them especially the United Nations Educational, Scientific & Cultural Organization (Paris), the United Nations Development

Programme (New York), the World Wide Fund for Nature (Gland, Switzerland), and the International Boundaries Research Unit (Durham, UK).

Consideration might be given to having the transfrontier reserve be designated as one or another of the following special categories: (a) a 'Biosphere Reserve' (Batisse 1986; IUCN 1990b: 241–253); (b) a 'World Natural Heritage' site (IUCN 1990b: 225–240) within the framework of the 1972 World Cultural & Natural Heritage Convention (UNTS 15511); or (c) a 'Wetland of International Importance' (IUCN 1990b: 257–275) within the framework of the 1971 Ramsar Convention on Wetlands of International Importance especially as Waterfowl Habitat (UNTS 14583). To pursue any of these three possibilities, it would be necessary to consult with the United Nations Educational, Scientific & Cultural Organization.

It need hardly be mentioned that each of the countries involved must have in place the necessary administrative infrastructure. If it does not already exist (or is inadequate), a department of national reserves should be created (or strengthened) within the ministry that deals with the environment or with natural resources in general. In this regard, a country could draw particularly upon advice from such agencies as the United Nations Development Programme or the International Union for Conservation of Nature.

The importance of building confidence as an approach to regional peace and security through such efforts as transfrontier collaboration is explained elsewhere (Goldblat 1993). Then there is the very sensitive issue of national sovereignty over natural resources, even when these are shared across national boundaries; and the growing recognition that transboundary environmental issues dilute the sovereign rights of states, at the same time imposing duties on them (Schrijver 1993). Nonetheless, it can be amply demonstrated that neighboring countries can and do collaborate successfully on a variety of often complex border-related issues (Blake 1993).

In many parts of the world the integrity of a protected natural area can be maintained most successfully with the active participation of the local rural population. The need for and benefits of such local participation to both the reserve and the local communities can be clearly delineated (Dennis and Spergel 1993; MacKinnon 1993; McNeely 1993). The range of approaches to protected-area management need to be examined (including the possibility of instituting a system of concentric zones), emphasizing the need to balance competing demands with a site-specific solution (McNeely 1993).

The formidable financial commitment that would be entailed by the countries in establishing a transfrontier reserve would in all likelihood have to be covered by outside sources. Here the countries might wish to turn to such agencies as the International Bank for Reconstruction & Development (World Bank) (Washington), the Global Environment Facility (Washington) (GEF 1992a), the World Wide Fund for Nature, or one of the regional United Nations development banks. Various possibilities for obtaining the necessary financial base to cover both initial capital expenses and continuing maintenance costs, favoring the establishment of a trust fund, are presented elsewhere (Dennis and Spergel 1993).

The possibility for establishing a transfrontier reserve for peace and nature in the Third World can be explored through the vehicle of an illustrative case study on the Indochinese peninsula. Such a case study might begin with a detailed examination of past conflict and cooperation within the region (Tønnesson 1993). This is done for the purpose of suggesting both the level of difficulty that might be associated in achieving the transfrontier cooperation being sought here and the great desirability of doing so. The actual process of establishing such a reserve in Indochina is presented elsewhere (MacKinnon 1993). Pointed out is not only the potential value of the proposed area for protecting regional biodiversity, further stressing that the first step on the path to creating a transfrontier reserve would be to establish separate, though coordinated, contiguous independent national reserves, a view that finds sympathy both in Cambodia (Mareth 1993) and Viet Nam (Hồe 1993). As another possibility, consideration might here be given for the proposed transfrontier reserve to be established under the aegis of the Committee for Coordination of Investigations of the Lower Mekong Basin (Bangkok). The proposed reserve falls within the watershed (catchment area; drainage basin) of the Mekong River; Laos and Viet Nam are members of the Committee and Cambodia may soon rejoin it[3]; and the Committee is in the process of integrating an environmental component into its watershed development projects (cf. MacKinnon 1993; Tønnesson 1993).

A draft (model) agreement for creating a transfrontier reserve, again employing Indochina as a concrete case in point is presented elsewhere (Westing 1993b). Offered at the same time is the presumably requisite precursor to such an agreement, a draft (model) interim memorandum of understanding for creating such a reserve that, *inter alia*, provides for the initial establishment of contiguous though independent national reserves. Background information for the illustrative Indochina case study is provided by a compilation of some relevant facts and figures (Westing 1993c) and by a reproduction of relevant past agreements among the Indochinese states (Westing 1993a).

6.5 Precedents

Precedents for the formal establishment of transfrontier reserves explored here address the two objectives envisioned for these reserves: those dealing with the protection of nature; and those dealing with the strengthening of peace and security, with special emphasis in the latter on demilitarization. Each category is reviewed in turn, but before doing so it should be of interest to note that the Consultative Assembly of the Council of Europe (Strasbourg, France) has recommended the creation of transfrontier protected natural areas as a means of strengthening the spirit of regional cooperation and solidarity (CoE 1970, Paragraph 6). Moreover,

[3] Cambodia rejoined the Committee in 1995 (which in addition to Laos and Viet Nam also included Thailand).

preparation of the necessary steps and legal instruments has been facilitated by a number of available sets of guidelines: for protected natural areas in general (Lausche 1980); for transfrontier cooperation in general (Dupuy 1982); and for transfrontier reserves in particular (Goetel 1964; IUCN 1990c; Nicol 1974).

6.5.1 Transfrontier Protected Natural Areas

The legitimacy of transfrontier protected natural areas has been established through a number of past actions, although no such reserve appears to have actually been established as yet. The multilateral treaties that provide their states parties with at least some legal basis for the formal creation of transfrontier protected natural areas include:

(a) the 1933 London Convention Relative to the Preservation of Flora and Fauna in their Natural State (LNTS 3995) (Kiss 1983: 57–64). This treaty commits the states parties to 'cooperation' with respect to contiguous protected natural areas (cf. Article 6);
(b) the 1971 Ramsar Convention on Wetlands of International Importance Especially as Waterfowl Habitat (UNTS 1458) (Kiss 1983: 246–248; Lyster 1985: 345–354; Weiss et al. 1992: 497–502). This treaty commits the states parties to 'consultation' with respect to a transfrontier wetland or water system (cf. Article 5);
(c) the 1979 Bern Convention on the Conservation of European Wildlife and Natural Habitats (UNTS 21159) (Kiss 1983: 509–519; Lyster 1985: 428–441). This treaty commits the states parties to 'coordination' in protecting natural habitats in frontier areas (cf. Article 4.4);
(d) the 1982 Protocol Concerning Mediterranean Specially Protected Areas (UNTS 24079) (Rummel–Bulska and Osafo 1991: 154–157; Weiss et al. 1992: 389–394). This treaty commits the states parties to 'consult' each other regarding a frontier protected area, and to examine the possibility of establishing a corresponding area (cf. Article 6); and
(e) the 1982 Benelux Convention on Nature Conservation and Landscape Protection (Brussels, 8 June 1982; in force, 1 October 1983; UNTS unlisted) (Rummel-Bulska and Osafo 1991: 163–164). This treaty commits the three states parties to develop a concept of transboundary natural areas and landscapes, to inventory them, to establish coordinate programs for their management and protection, and to seek their establishment (cf. Article 3).

Three additional treaties are considered to be of relevance here even though they do not deal specifically with transfrontier protected natural areas. The first is the 1973 Oslo Agreement on the Conservation of Polar Bears (Oslo, 15 November 1973; in force, 26 May 1976; UNTS unlisted), as a result of which all five of the arctic states encroaching upon the natural range of the polar bear (*Ursus maritimus* = *Thalarctos maritimus*; IUCN Vulnerable) have agreed to conserve this species, *inter alia*, by protecting the ecosystems of which it is a part (cf. Article 2) (Kiss 1983: 401–403; Lyster 1985: 407–410; Weiss et al. 1992: 486–488). The second is the 1974 Nordic Convention on the Protection of the Environment (UNTS 16770), as a result of which four Scandinavian states have agreed to cooperate in the mitigation of environmentally harmful transfrontier activities, in essence as if their national boundaries did not exist (Kiss 1983: 403–405; Weiss et al. 1992: 486–488). And the third is the 1980 European Convention on Transfrontier Cooperation (UNTS 20967), which facilitates and fosters cooperation by the states parties across their national frontiers (Rummel–Bulska and Osafo 1991: 93–106).

A number of bilateral treaties also exist, or are in various stages of consummation, that permit the establishment of coordinate protected natural areas on opposite sides of a shared border, noteworthy among them:

(a) The 1964 Treaty between Germany and Luxembourg for the Establishment of a Joint Nature Park (Clervaux [Clerf], Luxembourg, 17 April 1964; entry into force, 15 October 1965; UNTS unlisted), to be named the German-Luxembourgian Nature Park (43,000 ha in Germany, 36,000 ha in Luxembourg; established 1965; IUCN unlisted), by which the two states parties have agreed to establish mutually designated contiguous reserves enjoying equivalent levels of protection as well as a joint Advisory Commission;
(b) The 1971 Agreement between Belgium and Germany regarding Cooperation for the Establishment and Development of a Nature Park in the Regions of Nordeifel/Scheifel/Hohes Venn Eifel (Gemünd, Germany, 3 February 1971; in force, 3 February 1971; UNTS unlisted), by which the two states parties have agreed to establish mutually designated contiguous reserves enjoying equivalent levels of protection as well as a joint Advisory Commission; to include the Hautes Fagnes State Nature Reserve in Belgium (4,000 ha; established 1957; IUCN IV), the Hautes–Fagnes–Eifel Nature Park in Belgium (68,000 ha; established 1985; IUCN V), and the Nordeifel Nature Park in Germany (175,000 ha; established 1960; IUCN V);
(c) The 1976 Agreement between Germany and the Netherlands for Cooperation on the Establishment and Development of a Maas–Schwalm–Nette Nature Park (Düsseldorf, Germany, 30 March 1976; in force, 26 March 1977; UNTS unlisted), by which the two states parties have agreed to establish mutually designated contiguous reserves (the German reserve: 43,500 ha; established 1965; IUCN unlisted; the Netherlandic reserve: 22,000 ha; established 1976; IUCN unlisted) enjoying equivalent levels of protection as well as a joint Advisory Commission; and
(d) The 1989 Agreement between Finland and the USSR on the Friendship Nature Conservation Area (Helsinki, 26 October 1989; in force, 14 November 1990; UNTS unlisted), by which Finland has agreed to establish—via the merger of Ulvinsalo Strict Nature Reserve in Finland (2,500 ha; established 1956; IUCN I) with four other parcels—a 'Friendship Park' (20,400 ha) opposite

Kostomukhiskiy Nature Conservation Area in the Russian Federation (47,000 ha, to be enlarged to 48,000 ha; established 1983; IUCN I). The Agreement calls for the creation of a joint commission to guide cooperation between the two reserves via exchange of information, joint research programs, and other coordination. Protection, maintenance, and financing of the two reserves are to remain separate.

A number of other international agreements of relevance to the establishment of transfrontier reserves either exist or have been proposed (cf. Appendix 6.1). Moreover, various unilaterally managed pairs of contiguous protected natural areas exist in adjacent nations for which at least some level of informal coordination has been established (Saussay 1980; Thorsell and Harrison 1990). Several of the more prominent ones are enumerated here (IUCN 1990b):

(a) Waterton Lakes National Park in Canada (53,000 ha; established 1911; IUCN II; UNESCO biosphere reserve) opposite Glacier National Park in the USA (410,000 ha; established 1910; IUCN II; UNESCO biosphere reserve). They were linked symbolically in 1932 (cf. Appendix 6.2);
(b) Tatransky (High Tatra) National Park in Slovakia (74,000 ha; established 1948; IUCN II) opposite Tatra National Park in Poland (21,000 ha; established 1955; IUCN II). They were linked symbolically apparently in 1955 (cf. Appendix 6.3);
(c) Vanoise National Park in France (53,000 ha; established 1963; IUCN II) opposite Gran Paradiso National Park in Italy (70,000 ha; established 1922; IUCN II); it is an area of particular value for conserving the alpine ibex (*Capra ibex*; IUCN Least Concern). They were linked symbolically in 1972;
(d) Kluane National Park in Canada (2,200,000 ha; established 1972; IUCN II; world heritage) opposite Wrangell–Saint Elias National Park in the USA (3,400,000 ha; established 1978; IUCN II; world heritage). They were linked symbolically in 1979, at the time they were inscribed as 'World Natural Heritage' sites; and
(e) Mount Nimba Strict Nature Reserve in Guinea (13,000 ha; established 1944; IUCN I; world heritage; UNESCO biosphere reserve) opposite Mont Nimba Strict Nature Reserve in the Ivory Coast (5,000 ha; established 1944; IUCN I; world heritage). They were linked symbolically in 1982, at the time they were inscribed as 'World Natural Heritage' sites.

Finally, a strong possibility exists that a true transfrontier reserve will be established in the future that brings together the Russian Federation and the USA. Such a likelihood derives from a bilateral statement to establish the so-called 'Beringia International Park', by which the two states have agreed to establish a unified complex of terrestrial and aquatic protected areas (Bush and Gorbachev 1990). Although the modalities are yet to be formalized, it is expected that the reserve will be run under the bilateral authority of the US Secretary of the Interior and his or her Russian counterpart, with the advice of a joint Advisory

Commission. To be provided for in the process are eased travel restrictions, traditional uses by the native population, and research cooperation. The reserve is envisaged to include the Bering Land Bridge National Preserve in the USA (1,125,000 ha; established 1978; IUCN V).[4]

6.5.2 Demilitarized Border Regions

It is manifestly clear that zones being designated as protected natural areas deserve to be protected from disruption of all sorts, whether civil or military. Thus it is here suggested that the proposed reserve be formally demilitarized—a notion that should, in fact, be carried over to every protected natural area in the world. Adding to the value of this suggestion, is the recognition that demilitarization serves a useful role in political confidence building (Delbrück 1982).

The legitimacy of demilitarized border regions (and demilitarized oceanic islands) has been established through a number of past actions. The multilateral treaties that provide legal precedents or the legal basis for the formal creation of a demilitarized region include:

(a) the 1920 Spitsbergen Treaty (LNTS 41), which demilitarizes the Svalbard archipelago and Bear island in the Arctic Ocean (cf. Article 9) (Goldblat 1982: 132–133);
(b) the 1921 Aaland Island Convention (LNTS 255), which demilitarizes the Åland Islands in the Baltic Sea (cf. Articles 3 etc.) (Goldblat 1982: 133–134);
(c) the 1947 Italian Peace Treaty (UNTS 747), which demilitarizes a number of Italian islands (Pantellaria; the Pelagian Islands, including Lampedusa; and Pianosa) plus a number of now Greek islands (the Dodecanese islands, including Rhodes) in the Mediterranean Sea (cf. Articles 14 & 49) (Leiss and Dennett 1954: 163–250);
(d) the 1959 Antarctic Treaty (UNTS 5778), which demilitarizes the land area of Antarctica (cf. Article 1) (Goldblat, 1982: 150–153; Kiss 1983: 150–153; Weiss et al. 1992: 515–520); and
(e) the 1977 Protocol I on the Protection of Victims of International Armed Conflicts (UNTS 17512), which provides for the creation of demilitarized zones (cf. Article 60) (Goldblat 1982: 239–252).

Bilaterally demilitarized border regions include: (a) the Canadian/United States border, since 1817; (b) the Norwegian/Swedish border, since 1905; and (c) the Chinese/Nepalese border, since 1960.

---

[4] In January 2013 Russia established the Beringia National Park in its coastal Chutkutka Region, directly across from the US Bering Land Bridge Preserve, important for further setting the stage for bilateral cooperation.

6.6 Potential Sites

Of the 191 current nations in the world, about three-quarters share at least one border with another state, often sharing borders with two or more states. Moreover, national boundaries are often 'remote', sparsely populated, and relatively undisturbed areas. In fact, at least several hundred of the almost 7,000 protected natural areas in existence today abut (or come close to abutting) a national border. Of these extant frontier reserves, 100 or so—that is, 50 pairs or so—lie opposite each other (Thorsell and Harrison 1990). It is thus clear that many opportunities exist for neighboring countries to establish transfrontier reserves. Such joint endeavors could, of course, be based: (a) on existing pairs of adjacent frontier reserves; (b) on one existing frontier reserve and another newly created opposite it; or (c) on a pair of newly created contiguous frontier reserves.

A number of potential transfrontier reserves are suggested here that are deemed to have the potential for the two co-equal purposes being pursued here, that is, for: (a) the protection of nature (i.e., of biodiversity) in ecosystems that cross national boundaries on the one hand; and (b) the building of political confidence (i.e., of friendship) between neighboring states on the other. The examples are taken to illustrate diverse possibilities in various parts of the world. They involve disparate habitats as well as various political relationships and differing regimes of existing nature protection.

The list of proposed transfrontier protected natural areas is limited to the Third World, where the need for both nature protection and confidence building seems especially urgent. The list could readily be expanded, for example, by including some of the frontier reserves noted earlier, for example, the two contiguous protected natural areas in Guinea and the Ivory Coast involving Mount Nimba or the three in Benin, Burkina Faso, and Niger involving the Niger River watershed. At the same time it must be emphasized that the first proposal summarized below—referring to the Indochinese peninsula—has been selected for detailed examination in the present study.

6.6.1 Indochinese Peninsula

An 'Indochina tri-state reserve for peace and nature' would provide a mechanism for its three states—Cambodia, Laos, and Viet Nam (cf. Westing 1993c)—to deal with natural resources that are shared by three sovereign states, namely, a way for them to deal cooperatively with a common natural heritage of that ecogeographical region. Thus, the long-term goal for the Indochinese peninsula would be for the three states to maintain their sovereignty, while at the same time building confidence among themselves by dealing collaboratively with a shared natural resource (Findlay 1990; Tønnesson 1989; cf. Westing 1993a).

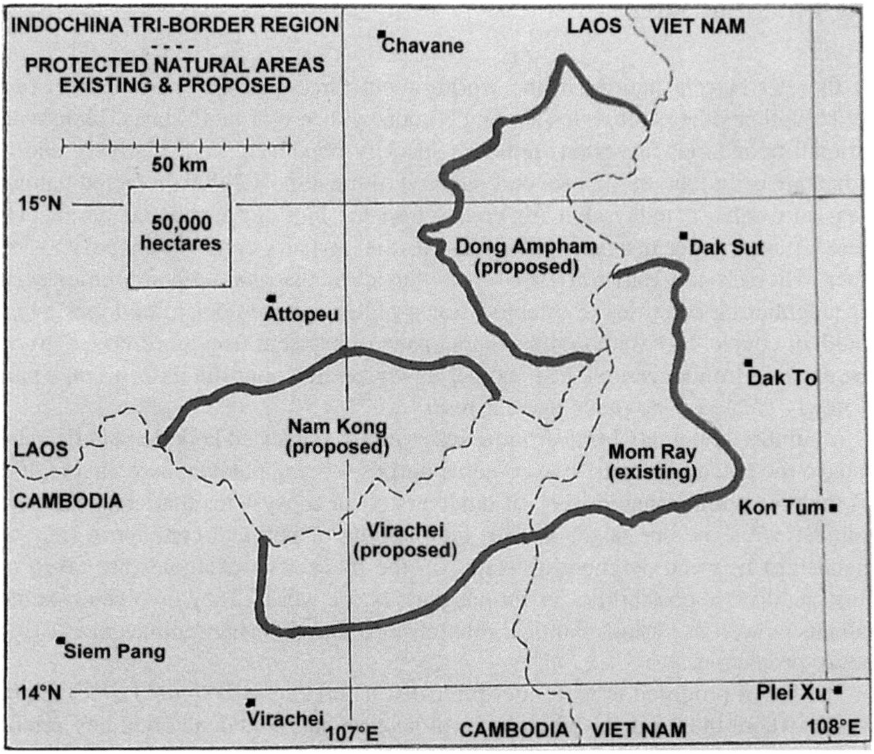

Fig. 6.1 Indochina tri-border region: protected natural areas, existing & proposed © Westing Associates in Environment, Security, & Education, by permission, 13 March 2013

The Indochinese reserve area would straddle all three countries, ca 500,000 ha in overall extent (with a minimum of 150,000 ha in each country). The area in question already includes the Mom Ray Nature Reserve in Viet Nam (45,000 ha, recently enlarged to 101,400 ha; established 1986; IUCN IV) as well as a number of proposed reserves in Cambodia and Laos (cf. Fig. 6.1). The area is rugged mountainous terrain, still relatively natural, being a sparsely populated mosaic of evergreen tropical montane forest and semi-deciduous or deciduous dipterocarp forest; the reserve area is the habitat of various animals threatened with extinction, including (IUCN 1990a; MacKinnon and Stuart 1989; Thouless 1987; Westing and Westing 1981): the kouprey (*Bos sauveli*; IUCN Critically Endangered), the red-shanked douc langur (*Pygathrix nemaeus*; IUCN Endangered), the green peafowl (*Pavo muticus*; IUCN Endangered), and some half dozen other large mammals and birds.

Whereas Viet Nam already has approximately 3 % of its area under nature protection, neither Cambodia nor Laos has as yet established such areas (IUCN 1990b: 56, 125, 206). Thus the proposed reserve would considerably bolster a presently most inadequate system of protected natural areas on the Indochinese peninsula (IUCN 1985). Moreover, the enabling trilateral agreement would provide a model for further cooperation among the three states; it would additionally provide a precedent for similar agreements elsewhere.

6.6.2 Korean Peninsula[5]

A 'Korean bi-state reserve for peace and nature' would, as above, provide a mechanism for its two states—the Democratic People's Republic of [North] Korea and the Republic of [South] Korea—to deal with natural resources that are shared by two sovereign states, namely, a way for them to deal cooperatively with a common natural heritage of that ecogeographical region. However, here the long-term goal for the Korean peninsula would be for the two states to become unified in due course, with the reserve serving as one of the steps to facilitate that process, while again at the same time dealing collaboratively with a shared natural resource (Cumings 1992; Sullivan and Foss 1987).

The Korean reserve might consist of two unconnected transfrontier zones, each a minimum of 50,000 ha in extent. Thus, there could be 25,000 ha or more on each side of the Military Demarcation Line in two distinct localities, thereby giving a combined over-all total of at least 100,000 ha. The fraction of such an area accounted for by the Demilitarized Zone (DMZ) would be roughly 20 %. For the first of the two areas, consideration should be given to a low wetland region about 60 km northeast of Panmunjom, important as a wintering ground for migratory birds, including at least two threatened with extinction (Archibald 1975; IUCN 1990a; Zimmerman 1981): the Japanese (red-crowned) crane (*Grus japonensis*; IUCN Endangered) and the white-naped crane (*Grus vipio*; IUCN Vulnerable). For the second of these areas consideration should be given to a mountainous temperate-forest region about 50 km southwest of the eastern terminus of the Demilitarized Zone.

Whereas South Korea already has approximately 6 % of its area under nature protection, only about 0.5 % of North Korea is similarly protected, the latter a truly inadequate fraction (IUCN 1990b: 124). A reserve area that straddles the Military Demarcation Line would be an especially suitable candidate for strict nature protection because the Demilitarized Zone—a strip of 4 km in width (2 km on either side of the Line)—has been left relatively undisturbed for almost the past four decades now. Moreover, South Korea maintains a 'Civilian Control Zone' of varying width (average, ca 5.4 km) south of the Demilitarized Zone that has also remained relatively undeveloped and to some extent undisturbed.

[5] The Korean Demilitarized Zone (DMZ) is the subject of Chap. 7.

6.6.3 Central Asian Mountains

A 'Mountain bi-state reserve for peace and nature' should as above, provide a mechanism for two of its states—China and Pakistan—to deal with natural resources that are shared by two sovereign states, namely, a way for them to cement their cordial relations by dealing cooperatively with a common natural heritage of that ecogeographical region (IUCN 1990d: 49).

China currently has only about 2 % of its area under nature protection, whereas approximately 5 % of Pakistan is similarly protected (IUCN 1990b: 68, 143). However, the Mountain reserve would join two existing reserves, Taxkorgan Nature Reserve in China (1,500,000 ha; established 1984; IUCN IV) with Khunjerab National Park in Pakistan (226,900 ha; established 1975; IUCN II). The area comprises the spectacularly beautiful junction of the Himalayan, Karakoram, and Pamir mountain ranges, containing some of the highest peaks in the world and many glaciers. Many of its large animals migrate seasonally across the national border. The area provides the habitat for a number of species threatened with extinction, including (IUCN 1990a; Schaller 1975–1977): the snow leopard (*Panthera uncia*; IUCN Endangered) and the Asiatic wild ass (*Equus hemionus*; IUCN Endangered). It also contains the locally threatened Marco Polo sheep (*Ovis ammon hodgsoni*; species, IUCN Near Threatened; subspecies, IUCN unlisted), coveted by trophy hunters for its huge horns.

6.6.4 Central American Isthmus

A 'Central American bi-state reserve for peace and nature' would, as above, provide a mechanism for two of its states—Costa Rica and Nicaragua—to deal with natural resources that are shared by two sovereign states, namely, a way for them to deal cooperatively with a common natural heritage of that ecogeographical region. The long-term goal for the reserve would be for the two states to maintain their sovereignty, while at the same time dealing collaboratively with a politically fractious area. The proposed reserve area—which on various past occasions has been considered by the two countries as a candidate for a transfrontier peace reserve (e.g., Guardian 1988; Karliner 1987)—has been the scene of much para-military activity (Garfield et al. 1987; Girot and Nietschmann 1991; Rice 1988–1989; Wille 1991). Moreover, Costa Rica is known for its pacific tendencies (Bird 1984; Westing 1990b); thus one hope for the proposed reserve would be for those tendencies to spread beyond its borders.

Whereas Costa Rica already has approximately 12 % of its area under nature protection, a most laudable fraction, a mere 0.3 % of Nicaragua is similarly protected, the latter far too little (IUCN 1990b: 77, 139). The Central American reserve area would encompass a substantial portion of the watershed of the San Juan River, which here straddles the as yet ill-defined boundary between the two countries, ca 400,000 ha in over-all extent. The area in question already includes the Barra del Colorado Wildlife Refuge in Costa Rica (92,000 ha; established 1985; IUCN IV). The area, which flanks the Caribbean Sea, is the largest remaining relatively intact tract of tropical rain forest on the isthmus. The habitat is a mosaic of mixed forests, swamps, lakes, and streams that are extremely rich in plant and animal species, both terrestrial and aquatic. It harbors various animals, including a host of threatened species, among them (IUCN 1990a): the West Indian manatee (*Trichechus manatus*; IUCN Vulnerable), Central American tapir (*Tapirus bairdii*; IUCN Endangered), jaguar (*Panthera onca*; IUCN Near Threatened), ocelot (*Leopardus pardalis* = *Felis pardalis*; IUCN Least Concern), and American crocodile (*Crocodylus acutus*; IUCN Vulnerable).

A bi-state reserve for peace and nature involving Costa Rica and Panama is another viable possibility (cf. Appendix 6.1.5).

6.6.5 Horn of Africa

An 'Ogaden bi-state reserve for peace and nature' would, again as above, provide a mechanism for two of its states—Ethiopia and Somalia—to deal with natural resources that are shared by two sovereign states, namely, a way for them to deal cooperatively with a common natural heritage of that ecogeographical region. Once again, the long-term goal for the reserve would be for the two states to maintain their sovereignty, while at the same time dealing collaboratively with a politically difficult area. The Ethiopian portion of the proposed area is part of a sparsely populated arid region of ca 25 million ha that has been claimed by Somalia since its independence in 1960. Fighting has occurred in the region on and off over the past three decades, with major armed conflicts in 1964, 1977–1979, 1982, and 1987 (East 1987; Woodward 1989).

The suggestion being offered here is that Ethiopia and Somalia each set aside a tract of perhaps 100,000 ha on either side of the *de facto* border. This is a region in which the nomadic patterns of the local people dependent upon livestock raising has traditionally resulted in transboundary movement. The area would be dedicated to peace and nature and be jointly administered, with the question of sovereignty being set aside in perpetuity. Such an agreement would serve to reinforce the peace settlement of sorts that was reached by the two countries in 1988. Indeed, the reserve might set an example to the rest of the Horn of Africa, now one of the most socially and environmentally insecure regions in the world (Hutchison 1991; Westing 1991).

Whereas Ethiopia already has approximately 5 % of its area under nature protection, *none* of Somalia is as yet similarly protected, the latter a manifestly inappropriate situation (IUCN 1990b: 85, 155). The proposed protected natural area supports (or has supported) a large number of species threatened with extinction, among them (IUCN 1990a; Largen and Yalden 1987; Lewis and Wilson 1975–1977): the hamadryas baboon (*Papio hamadryas*; IUCN Least

Concern), Grevy's zebra (*Equus grevyi*; IUCN Endangered), African elephant (*Loxodonta africana*; IUCN Vulnerable), Swayne's red hartebeest (*Alcelaphus buselaphus swaynei*; IUCN Endangered), Speke's gazelle (*Gazella spekei*; IUCN Endangered), and the scimitar-horned oryx (*Oryx dammah* = *O. tao*; IUCN Extinct in the Wild).

6.6.6 Northeast Africa

A 'Mount Elba region bi-state reserve for peace and nature' would provide a mechanism for its two states—Egypt and Sudan—to deal with natural resources that are shared by two sovereign states, namely, a way for them to deal cooperatively with a common natural heritage of that ecogeographical region. Such a reserve was endorsed in 1984 at a joint meeting of the Egyptian Academy of Scientific Research & Technology and the Sudanese National Council for Research, for purposes of strict nature protection, research, education, and tourism. The Egyptian portion would fall within a disputed zone of ca 1.8 million ha that formally became a part of Egypt in 1899, but which has been under the administrative control of Sudan since 1902 (Brownlie 1979: 110–120). The long-term goal would be to facilitate cooperation between the two states in the conservation and sustainable development not only of the proposed reserve, but as well of the disputed zone of which it would be a part. It would at the same time serve to reinforce the Brotherhood Charter concluded between Egypt and Sudan in 1987.

The proposed transfrontier reserve would span the eastern terminus of the official boundary between Egypt and Sudan—latitude 22°N—straddling that border between longitude 36°E and the Red Sea. It would encompass a contiguous area of perhaps 500,000 ha in each country, for a total of some 1 million ha. The region is sparsely populated with nomadic Beja pastoralists. It is a geographically and ecologically diverse region that contains a mountain range rising above 2,000 m (including Mount Elba in Egypt and Mount Asoteriba in Sudan) as well as lowland terrain and a varied coastline. The ecosystems include cloud forest (mist oasis) on the eastern slopes of the mountains, open woodland at their bases along the intermittent streams (wadis), desert, coastal mangrove, and salt marsh. Hundreds of plant and animal species have been recorded in the area. The area provides the habitat for a number of species threatened with extinction, including (IUCN 1990a): the African wild ass (*Equus africanus*; IUCN Critically Endangered), addax antelope (*Addax nasomaculatus*; IUCN Critically Endangered), Tora red hartebeest (*Alcelaphus buselaphus tora*; IUCN Critically Endangered), dama gazelle (*Nanger dama* = *Gazella dama*; IUCN Critically Endangered), and scimitar-horned oryx (*Oryx dammah* = *O. tao*; IUCN Extinct in the Wild).

Egypt currently has less than 1 % of its area under nature protection and Sudan only about 3 % (IUCN 1990b: 84, 165). However, the proposed transfrontier reserve would enhance only the modest amount now under protection in Sudan because Egypt's complementary contribution would encompass the already existing Gebel Elba Conservation Area in Egypt (480,000 ha; established 1986; IUCN IV).

6.7 Conclusion

From the foregoing examination, it appears that transfrontier (and perhaps also contested-island) protected natural areas as confidence-building measures might well turn out to have substantial value in strengthening comprehensive human security, especially so if emphasis for establishing such reserves were directed toward suitable sites in the Third World. It is especially the Third World that continues to experience violent border disputes; and it is especially the Third World that would benefit most from a building up of institutional awareness and institutional capability in the areas of environmental security on the one hand, and of political security on the other.

The establishment of a transfrontier protected natural area would have a number of beneficial outcomes. Such a reserve would contribute to the protection of an ever more seriously threatened global biosphere. An ecosystem or ecogeographical region, the integrity of which is compromised by straddling one or more national boundaries, would be safeguarded. Indigenous flora and fauna would be provided sanctuary, thereby contributing to the protection of global biodiversity (genetic resources). A transfrontier reserve would strengthen confidence, friendship, and cooperation between neighboring countries, at both their local and national levels. It might reduce the appalling worldwide frequency of border wars and other armed conflicts—and the death and destruction that invariably accompany them. Local communities would benefit, especially if the reserve were established taking both their rights and their sustainable development into consideration. It could be expected to lead to cross-border cooperation on other issues of mutual concern. And, finally, a transfrontier protected natural area would facilitate a recognition that notions of absolute state sovereignty must be relaxed somewhat to achieve human security on behalf of both present and future generations.

Appendix 6.1 Miscellaneous International Agreements (Actual or Proposed)

Appendix 6.1.1 Benin/Burkina Faso/Niger W Parks

A reserve established by France during the time it exerted control over West Africa (so named because the Niger River flowing through it takes the shape there of a 'W') subsequently became fragmented into three parts: W National Park in Benin (568,000 ha; established 1954; IUCN II); W National Park in Burkina Faso

(235,000 ha; established 1954; IUCN II); and W National Park in Niger (220,000 ha; established 1954; IUCN II). In 1965, the regional intergovernmental Council of the Entente (Abidjan) submitted a plan to the three relevant governments to establish a single reserve authority, which has not been acted upon as of early 1993 and which is no longer an activity of the Council.

Appendix 6.1.2 Roosevelt Campobello International Park

In 1964, Canada and the USA established the former Franklin Delano Roosevelt estate on Campobello Island (which is Canadian territory) as an 'International Park', placing it under the control of a joint Canadian–US 'International Park Commission' (UNTS 7674; for the relevant US legislation, cf. USCA 1992, Title 16, §§1101–1113). The Roosevelt Campobello International Park (1,000 ha; established 1964; IUCN unlisted), although not so much a natural reserve as an international memorial and museum, is of interest in the present context because it functions as a true bilateral entity under the legal control of a bilateral commission.

Appendix 6.1.3 Machias Seal Island International Park

Machias Seal Island is an approximately 6 ha island in the Bay of Fundy. It is located about 16 km from both Canada and the USA (at 44°30′N–67°08′W). Both countries claim the island (Line 1973; McNeil 1991). The island is an important nesting site for various colonial seabirds, including the Atlantic Puffin (*Fratercula arctica*; IUCN Least Concern), razorbill (*Alca torda*; IUCN Least Concern), Leach's storm petrel (*Oceanodroma leucorhoa*; IUCN Least Concern), and thin-billed murre (*Uria aalge*; IUCN Least Concern), and is also an important rest stop for numerous migratory birds—making the island a favorite site for ornithologists to visit. One solution to the contested sovereignty of the island would be to establish it as an international wildlife sanctuary, either under unchallenged or joint sovereignty (McNeil 1991).

Appendix 6.1.4 Bialowieza Primeval Forest

The Bialowieza forest, which straddles the Belarus/Polish border, is one of the largest remaining relatively undisturbed forest ecosystems in Europe, *inter alia*, providing the habitat for the wisent (*Bison bonasus*; IUCN Vulnerable) (Krasiúski 1990; Sokolowski 1983; Westing 1980: 57). Some level of protection is now provided by the Belovezhskaya Pushcha Hunting Reserve in Belarus (87,600 ha; established 1940; IUCN IV) opposite Bialowieza National Park in Poland

(5,300 ha; established 1932; IUCN II; world heritage; UNESCO biosphere reserve). The Global Environment Facility (Washington) (GEF 1992a) has recently approved a project, with substantial funding, in support of the Polish portion of the forest (GEF 1992b); and is considering similar support for the Belarus portion (GEF 1992b: 69–72).

Appendix 6.1.5 La Amistad International Park

Three contiguous protected natural areas encompass the middle and upper reaches of the Talamanca Mountain range, meant to protect one of the largest and most biologically diverse virgin forests in Central America: Chirripó National Park in Costa Rica (50,200 ha; established 1975; IUCN II), Cordillera de Talamanca (= La Amistad) National Park in Costa Rica (193,900 ha; established 1982; IUCN II; UNESCO biosphere reserve), and La Amistad National Park in Panama (207,000 ha; established 1988; IUCN II). Among the extraordinarily rich flora and fauna in these parks are a number of mammals threatened with extinction, including the ocelot (*Leopardus pardalis* = *Felis pardalis*; IUCN Least Concern), jaguar (*Panthera onca*; IUCN Near Threatened), and Central American tapir (*Tapirus bairdii*; IUCN Endangered); as well as a number of birds, including the resplendent quetzal (*Pharomachrus mocinno*; IUCN Near Threatened) and harpy eagle (*Harpia harpyja*; IUCN Near threatened). An agreement for natural resources border cooperation was concluded between Costa Rica and Panama in 1979, and a decision to establish an international friendship park was reached in 1982. Currently, the Organization of American States (Washington) is in the process of reaching an understanding with the United Nations Environment Programme (Nairobi) to assist Costa Rica and Panama in developing a program of forest management for the region, and to provide funding for that endeavor.

Appendix 6.2 Waterton–Glacier International Peace Park

In 1932, the US Congress authorized that, pending similar action by Canada, Glacier National Park 'shall become a part of an international park to be known as the Waterton–Glacier International Peace Park' (USCA 1992, Title 16, §161a). However, the portion within the USA 'shall be designated as the Glacier National Park' (USCA 1992, Title 16, §161b); and the Glacier National Park 'shall be under the exclusive control of the [US] Secretary of the Interior (USCA 1992, Title 16, §162). In the same year, the Canadian Parliament took similar unilateral action. However, no bilateral treaty has as yet been concluded between Canada and the USA that would provide for the establishment of a transfrontier protected natural area.

Appendix 6.3 The 1924 Cracow Protocol

In connection with a boundary dispute in the aftermath of World War I, representatives of Czechoslovakia and Poland produced a joint protocol (Cracow, 6 May 1924; entry into force, 5 September 1924; LNTS unlisted), *inter alia* recommending the establishment of four pairs of contiguous frontier protected natural areas and calling upon their two respective governments to conclude a bilateral treaty to enable such coordinate action (Goetel 1925; Goetel et al. 1926). The Protocol called for the establishment, in the first instance, of independent contiguous reserves in the border-straddling Tatra Mountains (outlined in the text above). Another of the three other recommendations has also come to pass: the Pieninsky National Park in Slovakia (2,000 ha; established 1967; IUCN II) opposite the Pieniny National Park in Poland (2,000 ha; established 1932; IUCN II). However, no bilateral treaty has as yet been concluded between Slovakia and Poland that would provide for the establishment of a transfrontier protected natural area.

References

Archibald, G. 1975. Cranes over Panmunjom. *International Wildlife* (Vienna, VA) 5(5):18–21.

Batisse, M. 1986. Developing and Focusing the Biosphere Reserve Concept. *Nature & Resources* (Paris) 22(3):1–10.

Bird, L. 1984. *Costa Rica: Unarmed Democracy*. London: Sheppard Press, 224 pp.

Blake, G.H. 1993. Transfrontier collaboration: a worldwide survey. In: Westing, A.H. (ed.). *Transfrontier Reserves for Peace and Nature: a Contribution to Human Security*. Nairobi: UN Environment Programme, 127 pp: pp 35–48.

Brownlie, I. 1979. *African Boundaries: a Legal and Diplomatic Encyclopædia*. London: C. Hurst, 1355 pp+2 maps.

Bush, G.H.W. & Gorbachev, M.S. 1990. Joint statement on the establishment of a Soviet–United States international park in the region of the Bering Strait, June 1, 1990. *Foreign Policy Bulletin* (Washington) 1(1):8–9.

CoE. 1970. *On the Creation of Regional and Transfrontier Natural Parks*. Strasbourg, France: Council of Europe, Consultative Assembly, Recommendation No. 587, 26 January 1970, 3 pp.

Cukwurah, A.O. 1967. *Settlement of Boundary Disputes in International Law*. Manchester, UK: Manchester University Press, 267 pp.

Cumings, B. 1992. Spring thaw for Korea's cold war? *Bulletin of the Atomic Scientists* (Chicago) 48(3):14–23.

Day, A.J. (ed.). 1987. *Border and Territorial Disputes*. 2nd edn. Harlow, UK: Longman, 462 pp.

Delbrück, J. 1982. Demilitarization, In: Bernhardt R. (ed.). *Encyclopedia of Public International Law. III*. Amsterdam: North Holland Publishing Company, 299 pp: pp 150–152.

Dennis, J.V. & Spergel, B.A. 1993. Protected natural areas: the financial challenge. In: Westing, A.H. (ed.). *Transfrontier Reserves for Peace and Nature: a Contribution to Human Security* Nairobi: UN Environment Programme, 127 pp: pp 59–65.

Downing, D. 1980. *Atlas of Territorial and Border Disputes*. London: New English Library, 121 pp.

Dupuy, P.-M. 1982. Legal aspects of transfrontier regional co-operation. *West European Politics* (London) 5(4):50–63.

East, R. 1987. Ethiopia–Somalia. In: Day A.J. (ed.). *Border and Territorial Disputes*. 2nd edn. Harlow, UK: Longman, 462 pp: pp 126–132.

Eckhardt, W. 1991. Wars and war-related deaths, 1500–1990. In: Sivard, R.L. (ed.). *World Military and Social Expenditures 1991*. 14th edn. Washington: World Priorities, 64 pp: pp 22–25.
Ehrlich, P. & Ehrlich, A. 1981. *Extinction: the Causes and Consequences of the Disappearance of Species*. New York: Random House, 306 pp.
Findlay, T. 1990. Stockholm on the Mekong?: CBMs for Asia/Pacific. *Pacific Review* (Oxford) 3:55–61.
Garfield, R.M. et al. 1987. Health-related outcomes of war in Nicaragua. *American Journal of Public Health* (Washington) 77:615–618.
GEF. 1992a. *Global Environment Facility: the Pilot Phase and Beyond*. Washington: International Bank for Reconstruction & Development (World Bank), GEF Working Paper Series No. 1, 52 pp.
GEF. 1992b. *Poland Forest Biodiversity Protection Project: Project Document*. Washington: International Bank for Reconstruction & Development (World Bank), 76 pp+2 maps.
Girot, P.O. & Nietschmann, B.Q. 1991. Transborder conflicts and cooperation: the geopolitics of the Nicaragua–Costa Rica borderland. *Boundary Bulletin* (Durham, UK) 1991(1):19–22.
Goetel, W. 1925. Great program of Poland and Czechoslovakia for national parks. *Zoological Society Bulletin* [now *Wildlife Conservation*] (New York) 28:26–36.
Goetel, W. 1964. Parks between countries. In: Adams, A.B. (ed.). *First World Conference on National Parks*. Washington: US National Park Service, 471 pp+17 pl.: pp 287–294.
Goetel, W. et al. 1926. [*Protocol: Conference of Polish and Czechoslovakian Scientists, Gathered in Cracow during 8 to 9 December 1925, on the Initiative of the Polish Academy of Science and the Czech Academy of Science*] (in Polish). Warsaw: Polish Academy of Science, 15 pp.
Goldblat, J. 1982. *Agreements for Arms Control: a Critical Survey*. London: Taylor & Francis, 387 pp.
Goldblat, J. 1993. Confidence building as an approach to regional peace and security. In: Westing, A.H. (ed.). *Transfrontier Reserves for Peace and Nature: a Contribution to Human Security*. Nairobi: UN Environment Programme, 127 pp: pp 17–20.
Grundy-Warr, C. (ed.). 1990. *International Boundaries and Boundary Conflict Resolution*. Durham, UK: University of Durham, International Boundaries Research Unit, 428 pp.
Guardian. 1988. Peace park. *Oryx* (London) 22:150–151.
Hoàng Hòe. 1993. An Indochina tri-state reserve: the view from Viet Nam. In: Westing, A.H. (ed.). *Transfrontier Reserves for Peace and Nature: a Contribution to Human Security*. Nairobi: UN Environment Programme, 127 pp: pp 93–98.
Hutchison, R.A. (ed.). 1991. *Fighting for Survival: Insecurity, People and the Environment in the Horn of Africa*. Gland, Switzerland: International Union for Conservation of Nature, 181 pp.
IUCN. 1985. *Corbett Action Plan for Protected Areas of the Indomalayan Realm*. Gland, Switzerland: International Union for Conservation of Nature, 23 pp.
IUCN. 1990a. *1990 IUCN Red List of Threatened Animals*. Gland, Switzerland: International Union for Conservation of Nature, 192 pp.
IUCN. 1990b. *1990 United Nations List of National Parks and Protected Areas*. Gland, Switzerland: International Union for Conservation of Nature, 275 pp.
IUCN. 1990c. Promoting effective management of transfrontier parks and reserves: guidelines. In: Thorsell, J. (ed.).*Parks on the Borderline: Experience in Transfrontier Conservation*. Gland, Switzerland: International Union for Conservation of Nature, 98 pp: pp 95–98.
IUCN. 1990d. *Resolutions and Recommendations: 18th Session of the General Assembly of IUCN — the World Conservation Union*. Gland, Switzerland: International Union for Conservation of Nature, 63 pp.
IUCN, UNEP, & WWF. 1991. *Caring for the Earth: a Strategy for Sustainable Living*. Gland, Switzerland: International Union for Conservation of Nature, 228 pp.
Karliner, J. 1987. Make parks, not war. *Amicus Journal* (New York) 9(4):8–13.
Kiss, A.C. (ed.). 1983. *Selected Multilateral Treaties in the Field of the Environment*. Nairobi: UN Environment Programme, 525 pp. [Cf. Rummel-Bulska & Osafo, 1991.]
Krasiúski, Z. 1990. Border where the bison roam. *Natural History* (NY) 1990(6):62–63.

Kratochwil, F. 1986–1987. Of systems, boundaries, and territoriality: an inquiry into the formation of the state system. *World Politics* (Princeton, NJ) 39:27–52.

Kratochwil, F. et al. 1985. *Peace and Disputed Sovereignty: Reflections on Conflict over Territory*. Lanham, MD: University Press of America, 159 pp.

Largen, M.J. & Yalden, D.W. 1987. Decline of elephant and black rhinoceros in Ethiopia. *Oryx* (London) 21:103–106.

Lausche, B.J. 1980. *Guidelines for Protected Areas Legislation*. Gland, Switzerland: International Union for Conservation of Nature, Environmental Policy & Law Paper No. 16, 108 pp.

Leiss, A.C. & Dennett, R. (eds). 1954. *European Peace Treaties after World War II: Negotiations and Texts of Treaties with Italy, Bulgaria, Hungary, Rumania, and Finland*. Boston: World Peace Foundation, 341 pp.

Lewis, J.G. & Wilson, R.T. 1975–1977. Plight of Swayne's hartebeest. *Oryx* (London) 13:490–494.

Line, L. 1973. Puffins without a country. *National Wildlife* (Vienna, VA) 11(2):36–39.

Lyster, S. 1985. *International Wildlife Law: an Analysis of International Treaties Concerned with the Conservation of Wildlife*. Cambridge, UK: Grotius Publications, 470 pp.

MacKinnon, J.R. 1993. An Indochina tri-state reserve: the practical challenges. In: Westing, A.H. (ed.). *Transfrontier Reserves for Peace and Nature: a Contribution to Human Security*. Nairobi: UN Environment Programme, 127 pp: pp 77–85.

MacKinnon, J.R. & Stuart, S.N. (eds). 1989. *The Kouprey: an Action Plan for its Conservation*. Gland, Switzerland: International Union for Conservation of Nature, 19 pp.

Mandel, R. 1980. Roots of the modern interstate border dispute. *Journal of Conflict Resolution* (Beverly Hills, CA) 24:427–454.

McNeely, J.A. 1993. Conservation and development: how protected natural areas can contribute to local communities. In: Westing, A.H. (ed.). *Transfrontier Reserves for Peace and Nature: a Contribution to Human Security*. Nairobi: UN Environment Programme, 127 pp: pp 49–58.

McNeely, J.A. et al. 1990. *Conserving the World's Biological Diversity*. Gland, Switzerland: International Union for Conservation of Nature et al., 193 pp.

McNeil, R.J. 1991. Machias Seal Island international park: a proposed resolution of a territorial conflict. In: Dykeman, F.W. (ed.). *Rural Land Management: Perspectives from Canada and the United States*. Sackville, New Brunswick: Mount Allison University, 168 pp: pp 91–102.

Mok Mareth. 1993. An Indochina tri-state reserve: the view from Cambodia. In: Westing, A.H. (ed.). *Transfrontier Reserves for Peace and Nature: a Contribution to Human Security*. Nairobi: UN Environment Programme, 127 pp: pp 87–92.

Myers, N. 1979. *The Sinking Ark: a New Look at the Problem of Disappearing Species*. Oxford: Pergamon Press, 307 pp.

Nicol, J.I. 1974. International parks and cooperation for more effective management. In: Elliott, H. (ed.). *Second World Conference on National Parks*. Gland, Switzerland: International Union for Conservation of Nature, 504 pp: pp 382–388.

Peters, P. et al. 1989. Responsibility of states in respect of the exercise of permanent sovereignty over natural resources. *Netherlands International Law Review* (Dordrecht) 36:285–313.

Pfeiffer, G., et al. 1982. *Comprehensive Study on Confidence-building Measures*. New York: UN Centre for Disarmament, Disarmament Study Series No. 7 (A/36/474), 42 pp.

Prescott, J.R.V. 1987. *Political Frontiers and Boundaries*. rev. edn. London: Allen & Unwin, 315 pp.

Reid, W.V. & Miller, K.R. 1989. *Keeping Options Alive: the Scientific Basis for Conserving Diversity*. Washington: World Resources Institute, 128 pp.

Rice, R.A. 1988–1989. Casualty of war: the Nicaraguan environment. *Technology Review* (Cambridge, MA) 92(4):62–71.

Rummel-Bulska, I. & Osafo, S. (eds). 1991. *Selected Multilateral Treaties in the Field of the Environment. II*. Cambridge, UK: Grotius Publications, 527 pp. [cf. Kiss, 1983.]

Saussay, C. du. 1980. Transfrontier parks. *Unasylva* (Rome) 32(127):16–22.

Schaller, G.B. 1975–1977. Mountain mammals in Pakistan. *Oryx* (London) 13:351–356.

Schrijver, N.J. 1988. Permanent sovereignty over natural resources versus common heritage of mankind: complementary or contradictory principles of international economic law? In:

Waart, P. de et al. (eds). *International Law and Development*. Dordrecht: Martinus Nijhoff, 457 pp: pp 87–101.

Schrijver, N.J. 1993. Sovereignty and the sharing of natural resources. In: Westing, A.H. (ed.). *Transfrontier Reserves for Peace and Nature: a Contribution to Human Security*. Nairobi: UN Environment Programme, 127 pp: pp 21–33.

Sharma, S.P. 1976. *International Boundary Disputes and International Law: a Policy-oriented Study*. Bombay: N.M. Tripathi, 323 pp.

Sokolowski, A.W. 1983. Restoring the bison's habitat in Bialowieza. *Ambio* (Stockholm) 12:197–202.

Starr, H. & Most, B.A. 1983. Contagion and border effects on contemporary African conflict. *Comparative Political Studies* (Beverly Hills, CA) 16:92–117.

Sullivan, J. & Foss, R. (eds). 1987. *Two Koreas — One Future?* Lanham, MD: University Press of America, 167 pp.

Tägil, S. 1982. The question of border regions in western Europe: an historical background. *West European Politics* (London) 5(4):18–33.

Tägil, S. et al. 1977. *Studying Boundary Conflicts: a Theoretical Framework* [transl. from the Swedish by M. Metcalf]. Lund: University of Lund, Lund Studies in International History No. 9, 215 pp.

Thorsell, J. & Harrison, J. 1990. Parks that promote peace: a global inventory of transfrontier nature reserves. In: Thorsell, J. (ed.). *Parks on the Borderline: Experience in Transfrontier Conservation*. Gland, Switzerland: International Union for Conservation of Nature, 98 pp: pp 3–21.

Thouless, C. 1987. Kampuchean wildlife: survival against the odds. *Oryx* (London) 21:223–228.

Tønnesson, S. 1989. Proposals for a lasting peace in Indochina. *Bulletin of Peace Proposals* [now *Security Dialogue*] (Oslo) 20:265–280.

Tønnesson, S. 1993. Peace and war: cooperation and conflict in 20th century Indochina. In: Westing, A.H. (ed.). *Transfrontier Reserves for Peace and Nature: a Contribution to Human Security*. Nairobi: UN Environment Programme, 127 pp: pp 67–76.

USCA. 1992. *United States Code Annotated*. St Paul, MN: West Publishing Company, numerous volumes & supplements [complete through 1991].

Vallentyne, J.R. & Beeton, A.M. 1988. 'Ecosystem' approach to managing human uses and abuses of natural resources in the Great Lakes basin. *Environmental Conservation* (Cambridge, UK) 15:58–62.

Weiss, E.B. et al. 1992. *International Environmental Law: Basic Instruments and References*. Ardsley-on-Hudson, NY: Transnational Publishers, 749 pp.

Westing, A.H. 1980. *Warfare in a Fragile World: Military Impact on the Human Environment*. London: Taylor & Francis, 249 pp.

Westing, A.H. 1982. War as a human endeavor: the high-fatality wars of the twentieth century. *Journal of Peace Research* (Oslo) 19:261–270.

Westing, A.H. 1984. How much damage can modem war create? In: Barnaby, F. (ed.). *Future War*. London: Michael Joseph, 192 pp: pp 114–124.

Westing, A.H. 1988. Military sector *vis-à-vis* the environment. *Journal of Peace Research* (Oslo) 25:257–264.

Westing, A.H. 1989a. Comprehensive human security and ecological realities. *Environmental Conservation* (Cambridge, UK) 16:295.

Westing, A.H. 1989b. Environmental approaches to regional security. In: Westing, A.H. (ed.). *Comprehensive Security for the Baltic: an Environmental Approach*. London: Sage Publications, 148 pp: pp 1–14.

Westing, A.H. 1989c. Environmental security for the Danube basin. *Environmental Conservation* (Cambridge, UK) 16:323–329.

Westing, A.H. 1990a. Our place in nature: reflections on the global carrying-capacity for humans. In: Polunin, N. & Burnett, J.H. (eds). *Maintenance of the Biosphere*. Edinburgh: Edinburgh University Press, 228 pp: pp 109–120.

Westing, A.H. 1990b. Towards eliminating war as an instrument of foreign policy. *Bulletin of Peace Proposals* [now *Security Dialogue*] (Oslo) 21:29–35.

Westing, A.H. 1991. Environmental security and its relation to Ethiopia and Sudan. *Ambio (Stockholm)* 20:168–171.

Westing, A.H. 1992. Protected natural areas and the military. *Environmental Conservation* (Cambridge, UK) 19:343–348.

Westing, A.H. 1993a. Existing regional Indochina agreements. In: Westing, A.H. (ed.). *Transfrontier Reserves for Peace and Nature: a Contribution to Human Security*. Nairobi: UN Environment Programme, 127 pp: pp 123–127.

Westing, A.H. 1993b. From hope to reality: establishing an Indochina tri-state reserve for peace and nature. In: Westing, A.H. (ed.). *Transfrontier Reserves for Peace and Nature: a Contribution to Human Security*. Nairobi: UN Environment Programme, 127 pp: pp 99–109.

Westing, A.H. 1993c. Indochinese peninsula: facts and figures. In: Westing, A.H. (ed.). *Transfrontier Reserves for Peace and Nature: a Contribution to Human Security*. Nairobi: UN Environment Programme, 127 pp: pp 119–121.

Westing, A.H. 1993d. Transfrontier reserves for peace and nature: select bibliography. In: Westing, A.H. (ed.). *Transfrontier Reserves for Peace and Nature: a Contribution to Human Security*. Nairobi: UN Environment Programme, 127 pp: pp 111–113.

Westing, A.H. & Westing, C.E. 1981. Endangered species and habitats of Viet Nam. *Environmental Conservation* (Cambridge, UK) 8:59–62.

Wille, C. 1991. Peace is hell! *Audubon* (NY) 93(1):62–70.

Willot, A. et al. 1981. *Study on All the Aspects of Regional Disarmament*. New York: UN Centre for Disarmament, Disarmament Study Series No. 3 (A/35/416), 59 pp.

Wolf, E.C. 1987. *On the Brink of Extinction: Conserving the Diversity of Life*. Washington: Worldwatch Institute, Paper No. 78, 54 pp.

Woodward, P. 1989. *War — or Peace — in North-east Africa?* London: Centre for Security & Conflict Studies, Conflict Study No. 219, 29 pp.

WRI. 1992. *World Resources 1992-93: a Report by the World Resources Institute*. 5th edn. New York: Oxford University Press, 385 pp.

Zimmerman, D.R. 1981. Fragile victory for beauty on an old Asian battleground. *Smithsonian* (Washington) 12(7):56–65.

Chapter 7
Regional Security: The Case of the Korean Demilitarized Zone (DMZ)

7.1 Three Nations Cut in Two

One of the momentous outcomes of World War II was that the German *Reich* was in 1945 cut in two, with about one-third of its territory and population forming East Germany (the German Democratic Republic), and the remainder becoming West Germany (the Federal Republic of Germany). As we all know, for the subsequent 35 years the two States went their remarkably separate ways politically, socially, and economically. And it is important for me to stress that right up to the 11th hour essentially no one foresaw the rapid collapse of both the physical and psychological barriers that had so firmly separated the two German States for so long. And I am pleased to be able to note that one outcome of reunification has been that there now exists a movement to convert the former fortified strip of land between the two Germanys into a nature-protected green belt.

I mention next (but for present purposes only in passing) that 1990 also witnessed the reunification of the two Yemens—and that merger occurring despite two decades of serious bilateral hostility, as well as of ideological and economic disparities comparable to those of the two Germanys. Thus, here again there occurred an unanticipated reunion, one that once initiated, progressed with

This chapter is reproduced from the author's Entry #441 provided in Chap. 2, with the title, '*The Korean Demilitarized Zone (DMZ) as a Bridge between the two Koreas*'. It was presented at 'A World without Walls 2010: an International Conference on Peace Building, Reconciliation and Globalization in an Interdependent World', Berlin, Germany, 6–10 November 2010, of the Institute for Cultural Diplomacy. The copyright was transferred from the Institute for Cultural Diplomacy to Westing Associates in Environment, Security, & Education on 17 March 2013, and is used here by permission of the latter, as given on that same date, 17 March 2013. A preliminary version had been presented at the IVth World Congress on National Parks and Protected Areas, Caracas, 10–21 February 1992, of the International Union for Conservation of Nature (Gland, Switzerland). The author thanks Carol E. Westing for constructive criticism. Moreover, he is pleased to acknowledge helpful suggestions on the MoU (Appendix 7.5) from George W. Archibald, Peter Hayes, L. Hall Healy Jr, Ke Chung Kim, and Jeffrey A. McNeely. All International Union for Conservation of Nature species threat categories have been updated to March 2013 (cf. www.iucnredlist.org).

A. H. Westing, *From Environmental to Comprehensive Security*, SpringerBriefs on Pioneers in Science and Practice 13,
DOI: 10.1007/978-3-319-00687-1_7, © The Author(s) 2013

lightning speed. So, as with Germany, it must have been some mix of ethnic, linguistic, cultural, and geographical ties that in the end overcame the huge existing asymmetries.

I now come to the remarkably similar story—the basis of this presentation—of yet another national division that occurred half way around the world only a few years after Germany's unfortunate dissection. Thus, one of the momentous outcomes of World War II plus the Korean War was that the Korean State was in 1953 also cut in two, with about one-half of its territory and one-third of its population forming North Korea (the Democratic People's Republic of Korea), and the remainder becoming South Korea (the Republic of Korea) [Sequence of relevant international conferences: Cairo, December 1943; Yalta, April 1945; Moscow, September 1945; New York, November 1947; New York, June 1950; Panmunjom, July 1953]. Once again, the two States went their remarkably separate ways politically, socially, and economically. But here the two stories diverge, inasmuch as the Korean separation has for 57 long years now remained (with minor fluctuations) about as firm as ever, and thus with no credible inkling as yet of reunification. Indeed, the two sides (formally, North Korea until recently plus China *versus* the UN Command in the firm hands of the USA, now informally plus South Korea) continue to function separately within the framework of what in 1953 was meant to be a temporary Armistice Agreement, but as yet with no Peace Treaty in sight.[1]

7.2 The Korean Situation Today

My hope for the future of the Korean peninsula and its two still antagonistic States depends on progress in two main areas: *First*, to be able at last to witness a peaceful and mutually beneficial formal end to the war in which I myself had fought more than half a century ago while serving as a US Marine under United Nations command. And *second*, to be able to help insure the environmental integrity of that peninsula, without which the long-term health and well-being of its people and wildlife would be in certain jeopardy. Fortunately, those two areas of concern could in my view be fruitfully linked, in that progress with the latter one (i.e., progress with environmental conservation) could facilitate success with the former one (i.e., progress with political *rapprochement*).

---

[1] The temporary North/South relationship established by the 1953 Armistice Agreement (Panmunjom, 27 July 1953) consisted of the Democratic Republic of [North] Korea plus China *versus* the United Nations Command (UNC) under the leadership of the USA. With most of the UNC member states (ca 16 in number, although not including South Korea) having over the years withdrawn, on 7 November 1978 the USA established the Republic of [South] Korea/USA Combined Forces Command (ROK/USA CFC) under US command. The UNC continues to exist, but apparently essentially in name only. On 11 March 2013 North Korea unilaterally declared the 1953 Armistice Agreement to be 'invalid'. The significance and ramifications of that declaration are as yet obscure.

For a region such as the Korean peninsula to offer an appropriate home both for its human inhabitants (with their necessary crops, livestock, and civil infrastructure) and for as many as possible of the remaining native plants and animals, requires a combination of (1) the sensitive use of all those lands sequestered for agriculture, industry, transportation, and so forth, and (2) the setting aside of some fraction of the peninsula as protected areas for the native flora and fauna. The first of those two requirements—the sensitive use of all lands—is now only inadequately met in both the North and South, and will thus require substantial educational efforts, legislation, and enforcement, but is not the subject of this presentation. The second of those two concerns—the *de jure* protection of some areas as nature reserves (bio-sanctuaries)—is even more seriously deficient in both the North and South, and leads me to what is to follow.

The present paucity of protected habitats on the Korean peninsula has deprived the peoples of the region of the many subtle continuing benefits deriving from adequate expanses of natural areas, the so-called ecosystem services. Among those often overlooked benefits of natural areas I might especially mention: purification of water and air, amelioration of local climate, limiting of erosion and protection of watersheds, making available wild medicinal plants, offering tranquility and inspiration, providing opportunities both for scientific research and eco-tourism, and offering somewhat of a counter-balance to the escalating environmental adversities to be expected as global warming continues. This substantial Korean paucity of bio-sanctuaries has also inexorably led to at least some extinctions and to the likelihood that others will follow suit. Indeed, listed among the wildlife currently known to be in danger of extinction on the Korean peninsular, primarily for lack of adequate habitat, are at least 29 species of birds, 6 of mammals, and even 1 each of a salamander and a dragonfly (cf. Appendix 7.1).

The 1953 Armistice Agreement that ended the North/South hostilities established a Military Demarcation Line (MDL) between the two States (which, as it happens, I helped to survey in 1952) flanked by a Demilitarized Zone—the DMZ—a roughly east–west green belt that traverses a full range of habitats (saltwater and freshwater, lowland and wetland, upland and highland, grassland and woodland). The DMZ is 4 km [2.5 mile] wide and approximately 248 km [154 miles] long, thus occupying an area of about 992 km^2 [383 $miles^2$]. Found in the DMZ are about one-third of the peninsula's higher (vascular) plants, one-half of its terrestrial mammals, and at least one-fifth of its birds. Its ecological importance derives in significant part from traversing that wonderfully representative sample of most of the peninsula's diverse ecosystems, most of them in the DMZ now largely unmolested by human action for over half a century. This has permitted those diverse ecosystems to be well on their way to recovering naturally from their extraordinarily serious wartime and other prior human disruptions (cf. Fig. 7.1). And the wetlands among them provide crucial wintering grounds for such charismatic and imperilled birds as the cranes, egrets, ibises, spoonbills, and storks that annually migrate between the Korean peninsula and China, Russia, or Japan (cf. Appendix 7.1.1)—the basis for a recent proposal to establish a so-called Northeast Asian Biodiversity Corridor (cf. Appendix 7.2, *Nautilus*).

Fig. 7.1 White-naped Cranes (*Grus vipio*; IUCN Vulnerable) in the Central Highland's Cheorwon Basin, within the Civilian Control Zone (CCZ), with the adjacent Korean Demilitarized Zone (DMZ) wilderness in the background, taken by George W. Archibald on 15 December 2012 © International Crane Foundation, by permission, 22 March 2013

Thus the DMZ could become the centerpiece of any effort to work toward environmental sustainability for the peninsula. If the DMZ (or at least substantial portions of it; and perhaps together with some adjacent areas) were to be conserved in perpetuity it would serve the crucial function of helping to conserve the Korean peninsula's environment, at the same time serving as an inspiring memorial tribute to the many soldiers and civilians of both sides who had lost their lives during the hostilities. And, as is to be developed next, it is my hope that it could additionally represent a magnificent apolitical avenue and ultimate monument to peace between two presently most uneasy neighbors—and conceivably in time even to their reunification (an aim, I might add, that the two Koreas actually proclaimed in a joint communiqué of 4 July 1972 and reaffirmed on 13 December 1992 and once again on 4 October 2007).

7.3 Protecting the DMZ as a Confidence- and Security-Building Measure

The first governmental indication of interest in a DMZ-centered nature reserve occurred in early 1991 when North Korea approached the UN Secretary-General (Kofi A. Annan) to explore such a possibility, a like step that was taken very soon thereafter by South Korea. The task was given over to the Executive Director of the UN Environment Programme (Mostafa K. Tolba), a job that in turn was assigned to me. However, in short order I was to discover that whereas South Korea was maintaining its interest in this investigation, most regrettably North Korea soon (in 1992) drew back from it. The next relevant official statement came in December 1997 when the President of South Korea (Kim Young Sam), in addressing the UN General Assembly, specifically expressed his hope that the two Koreas would cooperate with each other to protect and preserve the DMZ in order to turn it into a zone of peace and ecological integrity. By contrast, North Korea's response to that initiative came about a year later (in August 1999) with the abrupt statement that existing political problems continued to prevent such a possibility. Gentle nudges from time to time from the *UN Environment Programme*, the *UN Development Programme*, and a number of nongovernmental organizations (most recently, from IUCN) could not break the stalemate (cf. Appendix 7.2).

A serious problem being faced here is that the DMZ had been created at war's end largely (ca 85 %) by 'temporarily' expropriating (confiscating) privately owned lands. Thus, as soon as the 1953 Korean Armistice Agreement gives way to a Peace Treaty, the DMZ will cease to exist and those private lands will have to revert to their rightful owners. And that problem is exacerbated by national expropriations of abutting northern and southern buffer zones, also kept largely undeveloped over the years. In fact, the so-called Civilian Control Zone (CCZ) to the south averages fully 5.4 km [3.4 miles] in width, for an area of ca 1,339 km^2 [ca 517 $miles^2$]. Another significant problem is the strongly competing interests to make use of the sequestered lands for immediate human demands such as agricultural and industrial expansion. So, unless provision is made for the establishment of substantial state-owned bio-sanctuaries in all or parts of the DMZ plus some of its abutting northern and southern buffer zones prior to the consummation of the as yet unanticipated Korean Peace Treaty, that opportunity might well be lost forever.

And that, of course, is why the completely unanticipated actions that occurred two decades ago in Germany (and in Yemen) are so disquieting in the present context.

Diplomatic relations between the two Koreas have vacillated over the years between uneasy and dismal. Any of the occasional potentially promising initiatives by one side or the other—whether in the arena of high politics or low—have not to date led to anything fruitful in the present context. Even South Korean President Kim Dae Jong's 'sunshine policy' toward North Korea in the 1990s (for which he was honored with the Nobel Peace Prize for 2000) led to no tangible environmental cooperation. Thus it is my conviction that now is the time for North and South Korea to take diplomatic advantage of the ever more urgent necessity to conserve the peninsula's environment. They would thus be doing so in a thoroughly non-provocative, apolitical, and mutually beneficial fashion. Such parallel actions would in total constitute a so-called confidence- and security-building measure serving to ease the existing tensions and animosities between the two.

Fortunately, a solid legal basis for cross-border environmental cooperation is already in place for the two Koreas in that both are states parties especially to four enabling multilateral treaties: the *1945 Charter of the United Nations* (UNTS unlisted); the *1972 World Heritage Convention* (UNTS 15511); the *1977 Protocol I on International Armed Conflicts* (UNTS 17512); and the *1992 Biological Diversity Convention* (UNTS 30619). Details of the specific relevance of those four universal legal instruments (as well as of two additional ones) are appended (cf. Appendix 7.3.1). Additionally appended are specific details pertaining to a number of further quite instructive universal, regional, and bilateral legal instruments of indirect relevance to such cooperation (cf. Appendix 7.3.2), of which the existing bilateral ones might well be of particular interest as models (cf. Appendix 7.3.2.3). And compilations are also provided of intergovernmental agencies and nongovernmental organizations that could be turned to for assistance in this matter (cf. Appendix 7.2), as well as of relevant publications (cf. Appendix 7.4).

7.4 The Next Steps

It should be clear by now that there is an urgent need for the two Koreas to initiate steps, both individually and in time jointly, to set in motion conversion of the present *de facto* DMZ nature reserve (or at least substantial portions of it—perhaps together with some relevant contiguous areas) into the *de jure* transfrontier reserve for peace and nature so important to the future environmental and societal security of the Korean Peninsula. As already suggested, an adequate number of the fundamental legal foundations is already in place (cf. Appendix 7.3.1). Thus it has become my strong conviction that now is the time for the two Koreas to consummate a ***Memorandum of Understanding*** (**MoU**) that would commit them to forthwith setting in motion steps to *unilaterally* establishing abutting protected areas on their respective sides of the Military Demarcation Line (MDL), with the notion in mind of ultimately joining them as transfrontier protected areas. To this end, I have prepared the draft of a very permissive and non-threatening model **MoU** for the two Koreas to consider and, of course, for them to revise as felt necessary (cf. Appendix 7.5).

In unilaterally establishing their protected areas within and adjacent to the DMZ, consideration should be given by the two Koreas that various of them could in time be nominated—as ecologically appropriate—to become joint border-straddling 'World Natural Heritages', 'UNESCO/MAB Biosphere Reserves', a 'UNEP Regional Yellow/West Sea' (this perhaps together with China), and 'Wetlands of International Importance' (although for this last possibility, both would have to become states parties to the *1971 Wetland [Ramsar] Convention* (UNTS 14583) (cf. Appendix 7.3.1).

As already suggested, the two Koreas are fortunate in being able to make use of a number of publications of direct relevance to the DMZ (cf. Appendix 7.4.1), and a further number exists that provide theoretical background and rather detailed guidance for the establishment of transfrontier reserves for peace and nature (cf. Appendix 7.4.2). And, as also noted earlier, they are additionally fortunate in being able to turn to a number of international agencies and nongovernmental organizations for guidance and support. On the one hand, these include especially UNEP, UNESCO, UNDP, FAO, UN-REDD, and GEF; and on the other, IUCN, WWF, the *International Crane Foundation*, the *Peace Parks Foundation*, and the *DMZ Forum* (cf. Appendix 7.2).

It should also be useful for the two Koreas to be reminded that the notion of transfrontier parks for peace and nature is by no means a new one. As far back as 1924 representatives of Poland and the former Czechoslovakia set in motion the establishment of two pairs of cooperating contiguous nature reserves (in the Tatra Mountains straddling the now Polish/Slovakian border) for the express purpose of rebuilding bilateral trust as an approach to settling a World War I border dispute. As another example, in 1999 Ecuador and Peru established a demilitarized transfrontier park (the Cordillera del Condor Peace Park) to celebrate the post-war settlement of a boundary dispute and to commemorate the soldiers of both sides who had fallen in their protracted border war. In fact, more than two dozen formal bilateral transfrontier reserves for peace and nature now exist around the world (cf. Appendix 7.3.2.3).

So, let us hope and trust that the herein suggested diplomatic confidence-building approach to ameliorating the half-century of fear and distrust between the two Koreas will in time lead to their reconciliation—and, thereby, also to a world with one less wall.

Appendix 7.1 Known Imperilled DMZ Wildlife

Appendix 7.1.1 Birds

Crane, Hooded (*Grus monacha*)—IUCN Vulnerable
Crane, Red-crowned (or Manchurian) (*Grus japonensis*)—IUCN Endangered
Crane, White-naped (or Grey) (*Grus vipio*)—IUCN Vulnerable (cf. Fig. 7.1)
Eagle, Imperial (*Aquila heliaca*)—IUCN Vulnerable
Eagle, Steller's sea (*Haliaeetus pelagicus*)—IUCN Vulnerable
Egret, Chinese (*Egretta eulophotes*)—IUCN Vulnerable
Goose, Swan (*Anser cygnoides*)—IUCN Vulnerable
Grassbird, Marsh (*Locustella pryeri = Megalurus pryeri*)—IUCN Near Threatened
Grasshopper-warbler, Pleske's (*Locustella pleskei*)—IUCN Vulnerable
Greenshank, Nordmann's (*Tringa guttifer*)—IUCN Endangered
Gull, Chinese black-headed (*Larus saundersi*)—IUCN Vulnerable
Ibis, Crested (*Nipponia nippon*), **if** any in the DMZ—IUCN Endangered
Sandpiper, Spoon-billed (*Eurynorhynchus pygmeus*)—IUCN Critically Endangered
Spoonbill, Black-faced (*Platalea minor*)—IUCN Endangered
Stork, Oriental (*Ciconia boyciana*)—IUCN Endangered

[These **14** or **15** birds represent 48 % or 52 % of the 29 known imperilled bird species on the Korean peninsula.]

Appendix 7.1.2 Mammals

Bear, Himalayan black (*Ursus thibetanus*), **if** any in the DMZ—IUCN Vulnerable
Deer, Chinese water (*Hydropotes enermis*)—IUCN Vulnerable
Deer, Siberian musk (*Moschus moschiferus*)—IUCN Vulnerable
Tiger (*Panthera tigris*), **if** any in the DMZ—IUCN Endangered
[These **2–4** mammals represent 33–67 % of the 6 known imperilled mammal species on the Korean peninsula.]

Notes:

(a) Imperilled species on the Korean peninsula, not known to inhabit the DMZ, include: 14 birds, 2 terrestrial mammals, 1 salamander, 1 dragonfly, 4 marine fish, 6 marine mammals, and 1 net coral (but no plants):

For a grand total of 19 DMZ + 18 other terrestrial + 11 marine = 48

(b) IUCN 2009 threat data used (Extinct, Extinct in the Wild, Critically Endangered, Endangered, Vulnerable, Near Threatened, Least Concern, [and unlisted]) from: www.iucnredlist.org [as here revised to March 2013].
(c) Presence in DMZ primarily from: Kim, K.-G. & Cho, D.-G. 2005. Status and ecological resource value of the Republic of Korea's De-militarized Zone. *Landscape & Ecological Engineering* (Tokyo) 1(1):3–15. (cf. p. 12.)

Appendix 7.2 Agencies and Organizations Mentioned in this Chapter

DMZ Forum
1471 Wilson Rd, East Meadow, NY 11554, USA. www.dmzforum.org
A nongovernmental organization (NGO), its origin dating to 1994, and formally established in 1997 for the purpose of transforming the Korean Demilitarized Zone (DMZ) from a symbol of war to a place of peace among humans and between humans and nature, with its biological and cultural resources preserved in perpetuity, thereby enriching current and future generations of all Koreans.

FAO: Food and Agriculture Organization of the United Nations
Viale delle Terme di Caracalla, I-00153 Rome, Italy. www.fao.org
An international agency, established in 1945, *inter alia*, to help developing countries modernize and improve agriculture, forestry, and fishery practices and ensure good nutrition for all.

GEF: Global Environment Facility
1818 H St, NW, Washington, DC 20433, USA. www.gefweb.org

A partnership of UNDP, UNEP, and the International Bank for Reconstruction & Development (World Bank), established in 1991 for the purpose of helping developing countries fund projects and programs that protect the global environment.

International Crane Foundation
PO Box 447, Baraboo, WI 53913, USA. www.savingcranes.org
A nongovernmental organization (NGO) established in 1973 for the purpose of committing to a future where all cranes are secure, a future where people cooperate to protect and restore wild populations and their ecosystems. It established a Korean DMZ Task Force in 2010.

IUCN: International Union for Conservation of Nature
Rue Mauverney 28, CH-1196 Gland, Switzerland. www.iucn.org
A nongovernmental organization (NGO) established in 1948, its membership open to individuals, nongovernmental organizations (NGOs), intergovernmental agencies, and governments, for the purpose of protecting and sustainably using the Earth's resources. IUCN's **Regional Office for Asia**: 63 Sukhumvit Rd Soi 39, Bangkok 10110, Thailand. The ***IUCN Red List*** can be accessed at: www.iucnredlist.org.

Korea Maritime Institute
Mapogu, Sangam Dong 1652, Seoul, Republic of Korea. www.kmi.re.kr/english
A nongovernmental organization (NGO) established in 1984 that specializes in shipping economics, marine policy and affairs, and fisheries. It has since 2005 supported a transfrontier Marine Peace Park that would include the western terminus of the DMZ.

Nautilus Institute for Security and Sustainability
2130 Fulton St, San Francisco, CA 94117, USA. www.nautilus.org
A nongovernmental organization (NGO) established in 1992 for the purpose of making it possible to build peace, create security, and restore sustainability for all people in our time. In 2010 it presented a detailed proposal for the establishment of a Northeast Asian Biodiversity Corridor, so important for the birds that overwinter especially in the DMZ and migrate to China, Russia, and Japan for the summer.

Peace Parks Foundation
PO Box 12743, Stellenbosch 7613, South Africa. www.peaceparks.org
A nongovernmental organization (NGO) established in 1997 for the purpose of facilitating the establishment of transfrontier conservation areas (peace parks), thereby supporting sustainable economic development, the conservation of biodiversity, and regional peace and stability.

UNDP: United Nations Development Programme
1 UN Plaza, New York, NY 10017, USA. www.undp.org
An intergovernmental agency established in 1965, as a program of the United Nations for the purpose of being the United Nations's global development network, advocating for change and connecting countries to knowledge, experience, and resources to help build a better life.

UNEP: United Nations Environment Programme
PO Box 30552, Nairobi 00100, Kenya. www.unep.org
An intergovernmental agency established in 1972, as a subsidiary organ of the UN General Assembly for the purpose of providing leadership and encouraging partnerships in caring for the environment by inspiring, informing, and enabling nations and peoples to improve their quality of life without compromising that of future generations. UNEP's **Regional Seas Programme** information at: www.unep.org/regionalseas.

UNESCO: United Nations Educational, Scientific and Cultural Organization
7, Place de Fontenoy, F-75352 Paris 07SP, France. www.unesco.org
An intergovernmental agency established in 1945, as a program of the United Nations for the purpose of constructing the defenses of peace in the minds of men, contributing to peace and security by promoting collaboration between peoples through education, science, culture, and communication, this mission rooted in recognition of the fundamental unity of all members of the human family, based on the values of universal respect for justice, the rule of law, human rights, and fundamental freedoms. UNESCO's **Man and the Biosphere Programme** information at: www.unesco.org/mab.

UN-REDD: United Nations Collaborative Programme on Reducing Emissions from Deforestation and Forest Degradation in Developing Countries
11–13 Chemin des Anémones, CH-1219 Châtelaine, Geneva, Switzerland. www.un-redd.org
A partnership of FAO, UNDP, and UNEP, established in 2008 for the purpose of helping developing countries to reduce emissions from deforestation and forest degradation, thus contributing to the global fight against climate change for a healthier, greener tomorrow.

WWF: World Wide Fund for Nature
Ave du Mont-Blanc 27, CH-1196 Gland, Switzerland. www.panda.org
A nongovernmental organization (NGO) established in 1961 for the purpose of stopping the degradation of the planet's natural environment and to build a future in which humans live in harmony with nature, doing so by conserving the world's biological diversity, ensuring that the use of renewable natural resources is sustainable, and promoting the reduction of pollution and wasteful consumption.

Appendix 7.3 Legal Foundations

Appendix 7.3.1 Universal Legal Instruments of Greatest Relevance

Charter of the United Nations and Statute of the International Court of Justice.

San Francisco, 26 June **1945**; in force 24 October 1945; depositary, United States (Washington); secretariat, UN Secretary-General (New York); UNTS unlisted; states parties as of October 2010, 192 (98 %) of all 195.

➜ Article 1 commits the states parties to maintain international peace, develop friendly relations among nations, achieve international cooperation in solving international problems of an economic, social, cultural, or humanitarian character, and be a center for harmonizing the actions of nations in the attainment of these common ends.
DPR of [North] Korea: A state party since 1991.
Rep of [South] Korea: A state party since 1991.
China: A state party since 1945.
Russia: A state party since 1945.
Japan: A state party since 1956.

Convention on Wetlands of International Importance especially as Waterfowl Habitat.
Ramsar, Iran, 2 February **1971**; in force, 21 December 1975; depositary, UNESCO (Paris); secretariat ('bureau'), *International Union for Conservation of Nature [IUCN]* (Gland, Switzerland); UNTS 14583; states parties as of October 2010, 160 (82 %) of all 195.

➜ Article 5 commits the states parties to consultation with respect to a transfrontier wetland or water system. Article 2 provides for the establishment of Wetlands of International Importance.
DPR of [North] Korea: **Not** a state party [through to May 2013].
Rep of [South] Korea: A state party since 1997.
China: A state party since 1992.
Russia: A state party since 1977.
Japan: A state party since 1980.

Convention concerning the Protection of the World Cultural and Natural Heritage.
Paris, 23 November **1972**; in force, 17 December 1975; depositary, UNESCO (Paris); secretariat UNESCO (Paris), utilizing the technical services of the *International Union for Conservation of Nature [IUCN]* (Gland, Switzerland) in reference to World Natural Heritages; UNTS 15511; states parties as of October 2010, 187 (96 %) of all 195.

➜ Article 6 commits the states parties not to take any deliberate measures which might damage, directly or indirectly, a World Natural Heritage of outstanding universal value situated on the territory of other states parties, recognizing that such heritage constitutes a World Heritage for whose protection it is the duty of the international community as a whole to cooperate. Article 3 (in conjunction with Article 2) provides for the establishment of World Natural Heritages.
DPR of [North] Korea: A state party since 1998.
Rep of [South] Korea: A state party since 1988.
China: A state party since 1995.
Russia: A state party since 1988.
Japan: A state party since 1992.

Protocol [I] Additional to the 1949 Geneva Conventions and Relating to the Protection of Victims of International Armed Conflicts.
Bern, 12 December **1977**; in force, 7 December 1978; depositary, Switzerland (Bern); secretariat, *International Committee of the Red Cross* (Geneva); UNTS 17512; states parties as of October 2010, 170 (87 %) of all 195.

➔ Article 60 provides to the states parties the opportunity to create demilitarized zones.
DPR of [North] Korea: A state party since 1988.
Rep of [South] Korea: A state party since 1982.
China: A state party since 1983.
Russia: A state party since 1989.
Japan: A state party since 2004.

Convention on Biological Diversity.
Rio de Janeiro, 5 June **1992**; in force, 29 December 1993; depositary, UN Secretary-General (New York); secretariat, UN Environment Programme (Montreal); UNTS 30619; states parties as of October 2010, 193 (99 %) of all 195.

➔ Article 8.a commits the states parties to establishing a system of protected natural areas; Article 5 to cooperating among themselves; and Article 3 to ensuring that activities within their jurisdiction or control do not cause damage to the environment of other states or of areas beyond the limits of national jurisdiction.
DPR of [North] Korea: A state party since 1994.
Rep of [South] Korea: A state party since 1994.
China: A state party since 1993.
Russia: A state party since 1995.
Japan: A state party since 1993.

Convention on the Prohibition of the Use, Stockpiling, Production and Transfer of Anti-personnel Mines and on their Destruction.
Ottawa, 3 December **1997**; in force, 1 March 1999; depositary, UN Secretary-General (New York); secretariat, Implementation Support Unit (Geneva); UNTS 35597; states parties as of October 2010, 156 (80 %) of all 195.

➔ Article 1 commits the states parties never under any circumstance to use anti-personnel mines and to ensure the destruction of all anti-personnel mines.
DPR of [North] Korea: **Not** a state party [through to May 2013].
Rep of [South] Korea: **Not** a state party [through to May 2013].
China: Not a state party [through to May 2013].
Russia: Not a state party [through to May 2013].
Japan: A state party since 1998.
[It should be noted here that Article 2 of the 27 July 1953 Korean Armistice Agreement requires the removal by both sides of any known mine fields and other hazards in the DMZ.]

[As a point of interest, a major reason stated by the USA for being alone among its NATO allies to not become a state party to this *Convention* is its felt need to use land mines in impeding a feared attack by North Korea on South Korea, a reason that would presumably evaporate at such time that reunification occurs.]

Appendix 7.3.2 Various Legal Instruments of Indirect Relevance

Appendix 7.3.2.1 Universal Supportive Instruments

Convention relative to the Preservation of Flora and Fauna in their Natural State.
London, 8 November **1933**; in force, 14 January 1936; depositary (and secretariat), the United Kingdom (London); LNTS 3995; states parties as of October 2010, 11 (6 %) of all 195.

➜ Article 6 commits the states parties to cooperation with respect to contiguous protected natural areas.
DPR of [North] Korea: Not a state party.
Rep of [South] Korea: Not a state party.
China: Not a state party.
Russia: Not a state party.
Japan: Not a state party.

Convention on the Conservation of Migratory Species of Wild Animals.
Bonn, 23 June **1979**; in force, 1 November 1983; depositary, Germany (Bonn); secretariat, UN Environment Programme (Bonn); UNTS 28395; states parties as of October 2010, 114 (58 %) of all 195.

➜ The treaty provides for the protection of wild animals that migrate across or outside national boundaries.
DPR of [North] Korea: Not a state party.
Rep of [South] Korea: Not a state party.
China: Not a state party.
Russia: Not a state party.
Japan: Not a state party.

Appendix 7.3.2.2 Regional Supportive Instruments

Scandinavian Convention on the Protection of the Environment.
Stockholm, 19 February **1974**; in force, 5 October 1976; depositary (and secretariat), Sweden (Stockholm); UNTS 16770; states parties as of October 2010, 4 (100 %) of 4.

➔ The treaty commits the states parties to cooperate in the mitigation of environmentally harmful transfrontier activities, in essence as if their national boundaries did not exist.

European Convention on the Conservation of Wildlife and Natural Habitats.
Bern, 19 September **1979**; in force, 1 June 1982; depositary (and secretariat), Council of Europe (Strasbourg, France); UNTS 21159; states parties as of October 2010, 47 (92 %) of Europe's 51 (plus 4 African states parties).

➔ Article 4.4 commits the states parties to coordination in protecting natural habitats in frontier areas.

European Convention on Transfrontier Co-operation.
Madrid, 21 May **1980**; in force, 22 December 1981; depositary (and secretariat), Council of Europe (Strasbourg, France); UNTS 20967; states parties as of October 2010, 36 (71 %) of Europe's 51.

➔ The treaty commits the states parties to facilitate and foster cooperation across their national frontiers.

Mediterranean Protocol concerning Specially Protected Areas.
Geneva, 3 April **1982**; in force, 23 March 1986; depositary, Spain (Madrid); secretariat, UN Environment Programme (Athens); UNTS 24079; states parties as of October 2010, 21 (100 %) of 21.
[This instrument is a protocol to the 1976 Barcelona Convention for the Protection of the Mediterranean Sea Against Pollution (UNTS 16908).]

➔ Article 6 commits the states parties to consult each other regarding a frontier protected area, and to examine the possibility of establishing a corresponding area.

Benelux Convention on Nature Conservation and Landscape Protection.
Brussels, 8 June **1982**; in force, 1 October 1983; depositary, Benelux Economic Union (Brussels); UNTS unlisted; states parties as of October 2010, 3 (100 %) of 3.

➔ Article 3 commits the states parties to develop a concept of transboundary natural areas and landscapes, to inventory them, to establish coordinate programs for their management and protection, and to seek their establishment.

European Convention on the Protection and Use of Transboundary Watercourses and International Lakes.
Helsinki, 17 March **1992**; in force, 6 October 1996; depositary, UN Secretary-General (New York); secretariat, UN Economic Commission for Europe (Geneva); UNTS 33207; states parties as of October 2010, 38 (75 %) of Europe's 51.

➔ Article 2 commits the states parties to ensure that transboundary waters are used with the aim of ecologically sound, rational, and equitable management, and to take measures for the prevention, control, and reduction of water pollution.

Convention on the Protection of the Marine Environment of the Baltic Sea Area.
Helsinki, 9 April **1992**; in force, 17 January 2000; depositary, Finland (Helsinki); secretariat, Finland via the 'Helsinki Commission' or HELCOM (Helsinki);

UNTS 36495; states parties as of October 2010, 9 (100 %) of the 9 littoral states (plus the European Economic Community).

[This instrument supercedes the 1974 Convention of same name (UNTS 25986).]

➜ The treaty provides a most useful model for establishing a UNEP Regional Sea.

Appendix 7.3.2.3 Bilateral Supportive Instruments

Agreement between Canada and the United States of America relating to the Establishment of the Roosevelt Campobello International Park.

Washington, 22 January **1964**; in force, 14 August 1964; UNTS 7674; states parties as of October 2010, 2 (100 %) of 2.

➜ The treaty commits the states parties to establishing an international memorial park under the control of a Joint Canadian-United States International Park Commission, i.e., functioning as a true bilateral entity under the legal control of a bilateral commission.

Treaty between Germany and Luxembourg for the Establishment of a Joint Nature Park.

Clervaux (Clerf), Luxembourg, 17 April **1964**; in force, 15 October 1965; UNTS unlisted; states parties as of October 2010, 2 (100 %) of 2.

➜ The treaty commits the states parties to establish mutually designated contiguous reserves enjoying equivalent levels of protection, as well as a Joint Advisory Commission.

Agreement between Belgium and Germany regarding Cooperation for the Establishment and Development of a Nature Park.

Gemünd, Germany, 3 February **1971**; in force, 3 February 1971; UNTS unlisted; states parties as of October 2010, 2 (100 %) of 2.

➜ The treaty commits the states parties to establish mutually designated contiguous reserves enjoying equivalent levels of protection, as well as a Joint Advisory Commission.

Agreement between Germany and the Netherlands for Cooperation on the Establishment of a Nature Park.

Düsseldorf, Germany, 30 March **1976**; in force, 26 January 1977; UNTS unlisted; states parties as of October 2010, 2 (100 %) of 2.

➜ The treaty commits the states parties to establish mutually designated contiguous reserves enjoying equivalent levels of protection, as well as a Joint Advisory Commission.

[Joint Declaration (between Costa Rica and Panama) over La Amistad Park] (In Spanish).

Guabito, Panama, 3 March **1979**; in force, 6 September 1988; UNTS unlisted; states parties as of October 2010, 2 (100 %) of 2.

➜ The treaty commits the states parties to establish contiguous national reserves, and to cooperating via a Bi-national Technical Commission.

[This transfrontier endeavor has a convoluted history. Respective Executive decrees in May 1982 created *La Amistad [= Friendship] International Park.* The Costa Rican and Panamanian portions together became a UNESCO/MAB Biosphere Reserve in 1982 (the Panamanian portion confirmed in 2000). The two national portions together became a transboundary World Natural Heritage in 1983. The original Agreement of 3 March 1979 was confirmed by Costa Rica in February 1982, but could not enter into force until Panama did so as well, on 6 September 1988. Subsequently, the Presidents of Costa Rica and Panama met in Sixaola, Costa Rica on 3 May 1992 to sign an Agreement for generalized cooperation in frontier development. The originally called for La Amistad Bi-national Technical Commission was finally created by a joint Agreement on 23 January 1996, which then functioned for some years before becoming inactive.]

Agreement between Finland and Russia on the Friendship Nature Conservation Area.

Helsinki, 26 October **1989**; in force, 14 November 1990; UNTS unlisted; states parties as of October 2010, 2 (100 %) of 2.

➜ The treaty commits the states parties to establish contiguous ‘Friendship Parks’, as well as a Joint Commission to provide cooperation via exchange of information, joint research programs, and other coordination, but with protection, maintenance, and financing to remain separate.

[Global and Definitive Peace Agreement between Ecuador and Peru] (in Spanish).

Brasilia, 26 October **1998**; in force, 26 October 1998; UNTS unlisted; states parties as of October 2010, 2 (100 %) of 2 (plus 4 guarantor states).

➜ Article 7 commits the states parties to create two contiguous environmental protection areas, but which remain under the sovereignty and jurisdiction of the two respective states.

[These two protection areas are to be together known as the *Cordillera del Condor Peace Park* and to serve to commemorate the soldiers on both sides who had fallen in the war.]

Appendix 7.4 Select Bibliography (References)

Appendix 7.4.1 Of Direct Relevance to the DMZ

Archibald, G.W. 1975. Cranes over Panmunjom: how Korea’s demilitarized zone became a lush wildlife sanctuary. *International Wildlife* (Reston, VA) 5(5):18–21.

Brady, L.M. 2008. Life in the DMZ: turning a diplomatic failure into an environmental success. *Diplomatic History* (Malden, MA) 32(4):585–611.

Healy, H. 2007. Korean demilitarized zone: peace and nature park. *International Journal on World Peace* (St Paul, MN) 24(4):61–83.

Higuchi, H. & Minton, J. 2000. The importance of the Korean DMZ to threatened crane species in northeast Asia. *Global Environmental Research* (Tokyo) 4(2):123–132.

Kim, K.C. 1997. Preserving biodiversity in Korea's demilitarized zone. *Science* (Washington) 278(5336):242–243.

Kim, K.C. 2007. Preserving Korea's demilitarized corridor for conservation: a green approach to conflict resolution. In: Ali, S.H. (ed.). *Peace parks: conservation and conflict resolution.* Cambridge, MA: MIT Press, 406 pp: pp 239–259 + *passim* 343–377.

Kim, K.-G. & Cho, D.-G. 2005. Status and ecological resource value of the Republic of Korea's De-militarized Zone. *Landscape & Ecological Engineering* (Tokyo) 1(1):3–15.

Shore, W.B. 2004. Sanctuary for nature and the dead: preserving the Korean Demilitarized Zone. *World Watch* (Washington) 17(6):32–35.

Westing, A.H. 1998–1999. A transfrontier reserve for peace and nature on the Korean peninsula. *International Environmental Affairs* (Hanover, NH) 10(1):8–17.

[Similar in: *Environmental Awareness* (Vadodara, India) 23(3):85–99. 2000.]

Westing, A.H. 2001. A Korean DMZ park for peace and nature: towards a code of conduct. In: Kim, C.–H. (ed.). *The Korean DMZ: reverting beyond division.* Seoul: Sowha Publishing Company, 359 pp: pp 157–191.

Westing, A.H. 2010. Toward environmental sustainability and reduced tensions on the Korean peninsula. *Environment* (Philadelphia, PA) 52(1):20–23.

Appendix 7.4.2 Of Assistance in Establishing a Transfrontier Reserve

Ali, S. (ed.). 2007. *Peace Parks: Conservation and Conflict Resolution.* Cambridge, MA: MIT Press, 406 pp.

[Primarily useful for conceptual background. Chapter 13, by K.C. Kim, provides an analysis of conserving the Korean DMZ as a 'green' approach to conflict resolution.]

Alsdirawi, F. & Faraj, M. 2004. Establishing a transboundary peace park in the demilitarised zone (DMZ) on the Kuwaiti/Iraqi borders. *Parks* (Gland, Switzerland) 14(1):48–55.

[Interesting parallel to the Korean situation.]

Oszlanyi, J. et al. 2000. *Recommendations for the Establishment and Functioning of Transboundary Biosphere Reserves.* Paris: UNESCO, MAB Report Series No. 69, 192 pp: pp 55–57.

[Useful for specific guidelines.]

Sandwith, T. et al. 2001. *Transboundary Protected Areas for Peace and Co–operation*. Gland, Switzerland: *International Union for Conservation of Nature* [IUCN], 111 pp.

[Useful for both theory and specific guidelines.]

United Nations. 1982. *World Charter for Nature*. New York: UN General Assembly Resolution No. 37/7, 28 October 1982, 5 pp.

[Useful as an overall conceptual and operational framework, and with Article 21 offering some specific guidance. 111 of the 156 UN members of the time (including China, Russia, and Japan) voted in favor; and at least 3 further states later formally advised the UN that they too supported the Charter. Neither Korea was a UN member at that time, but could now, of course, also endorse this epochal document.]

Westing, A.H. (ed.). 1993. *Transfrontier Reserves for Peace and Nature: a Contribution to Human Security*. Nairobi: UN Environment Programme [UNEP], 127 pp.

[Useful for both theory and specific guidelines.]

Appendix 7.5 The Memorandum of Understanding (MoU)

MEMORANDUM OF UNDERSTANDING (MoU) BETWEEN THE DEMOCRATIC PEOPLE'S REPUBLIC OF KOREA (DPRK) AND THE REPUBLIC OF KOREA (ROK) REGARDING FUTURE PROTECTION OF BOTH NATURE AND CULTURE IN THE DEMILITARIZED ZONE (DMZ) PLUS ITS CONTIGUOUS NORTHERN BUFFER ZONE (NBZ) AND SOUTHERN BUFFER (CIVILIAN CONTROL) ZONE (SBZ)

PREAMBLE

I. Recalling our commitment via Article 1 of the 1945 *Charter of the United Nations and Statute of the International Court of Justice* to maintain international peace, develop friendly relations among nations, and achieve international cooperation; and furthermore

II. Mindful of our commitment via Article 8.a of the 1992 *Convention on Biological Diversity* (UNTS 30619) to establish a system of protected natural areas; and via Article 5 to cooperate with other nations to that end; and furthermore

III. Recognizing the commitment by at least one of us (ROK) via Article 2 of the 1971 *Convention on Wetlands of International Importance especially as Waterfowl Habitat* (UNTS 14583) to establish wetlands of international importance; and via Article 5 to consultation with respect to a transfrontier wetland or water system; and furthermore

IV. Noting our affinity with the 1982 *World Charter for Nature* (UNGA Res 37/7) in providing an overall conceptual framework for our relationship with the natural world; and in particular with the general guidelines for cooperation offered by Article 21; and furthermore

V. Understanding the obligation of all nations not only to respect the whole of nature within and beyond their national domains, but more specifically to also protect in perpetuity some fraction of their own flora, fauna, and associated habitats—doing so both on behalf of the biota *per se* and in order to ensure the long-term survival and well-being of their own human inhabitants; and, moreover, realizing that such obligation to protect nature should in principle be independent of any unrelated political considerations; and furthermore

VI. Knowing that the DMZ and its contiguous NBZ and SBZ have to a major extent recovered their ecological integrity on a *de facto* basis over the past half-century or so, thereby providing a priceless *de jure* opportunity for us to add to our as yet modest amounts of necessary habitat allocated in perpetuity for nature; in the additional recognition that they hold irreplaceable natural habitats flourishing with native species of plants and animals already lost elsewhere on the peninsula, thus providing a critical resource for the peninsula's nature restoration and conservation; and also aware of the archeological, historical, spiritual, recreational, and similar cultural inclusions that would be simultaneously protected; and furthermore

VII. Recalling our long-term concerns over the environmental and cultural future of the DMZ, as exemplified, *inter alia*, by the individual approaches we each made in 1991 to the Secretary General of the United Nations to explore the possibility of a DMZ-centered reserve for peace and nature; and furthermore

VIII. In sympathy with the 2008 United Nations Collaborative Programme of FAO, UNDP, and UNEP on Reducing Emissions from Deforestation and Forest Degradation in Developing Countries (UN-REDD), as well with the *2009 United Nations Framework Convention on Climate Change* (UNTS 30822) draft policy approaches and positive incentives for that Programme (REDD +), which, *inter alia*, emphasize the role of conservation and enhancement of forest carbon stocks; and furthermore

IX. In the knowledge that encouragement, support, and technical assistance in this endeavor would be available from UNEP, UNDP, FAO, UN-REDD, UNESCO, IUCN, the *International Crane Foundation*, the *Peace Parks Foundation*, the *DMZ Forum,* and other international agencies and nongovernmental organizations; and in the further knowledge of a number of precedents elsewhere in the world; we now therefore enter into the following bilateral understanding:

OUR UNDERSTANDING

Article 1. We express our firm desire to protect in perpetuity as much as possible of the DMZ, its contiguous NBZ and SBZ, and its two associated coastal sea areas as a so-called green belt across our peninsula, expressing this desire on behalf of both nature and our citizenry, the latter owing to the several crucial so–called ecosystem services provided as well as to the protection of archeological, historical, spiritual, recreational, and similar cultural inclusions variously supportive, *inter alia*, of science, education, and tourism; and, moreover, for such green belt to serve as an inspiring permanent living memorial to all those who lost their lives in the Korean War of 1950–1953.

Article 2. We express our intention at some early date to unilaterally identify one or more sites within our half of the DMZ and its contiguous NBZ or SBZ, each of perhaps ten-thousand (10,000) ha or more in size, and each consisting of a natural, or natural plus cultural, area worthy of protecting in perpetuity as a reserve. It is our intention to choose each site on the basis of its value in protecting some special habitat (upland, wetland, grassland, woodland, mountain, plain, coastal sea, etc.) as well as of the richness or uniqueness of the biological diversity (biodiversity) of its indigenous flora and fauna, including the number of supported biota threatened with extinction, among the latter, e.g., the red-crowned crane (*Grus japonensis*; IUCN Endangered), so widely revered on the Korean peninsula as a symbol of peace, prosperity, and long life; and also, as appropriate, to protect any archeological, historical, spiritual, recreational, or similar cultural inclusions of lasting importance. Some of the potential sites it is our intention to consider are presented in *Annex 1* [Appendix 7.5.1].

Article 3. We express our intention to designate each such site described in Article 2 as a protected area enjoying a level of protection equal to one or another of the various IUCN Protected Area Management Categories, as described in *Annex 2* [Appendix 7.5.2]. It is our intention to forthwith set in motion establishment of any portion of the site falling within our own NBZ or SBZ, and to similarly provide forthwith for the establishment of any portion falling within our half of the DMZ as soon as that becomes legally possible.

Article 4. We express our intention that to the extent that a site chosen by the DPRK is contiguous with a site chosen by the ROK along the boundary between the two Koreas (along the Military Demarcation Line), both the DPRK and the ROK agree in time to a direct exchange of information between the contiguous two local protected area authorities on purely technical matters, e.g., on transboundary wildfire control, transboundary movement of wildlife, and transboundary floral or faunal disease or pest control.

Article 5. We express our intention at an early date to consider to set in motion—as ecologically appropriate—the necessary steps to designate any protected area we unilaterally establish within our half of the DMZ and its contiguous NBZ or SBZ as one or more of the following: (a) a 'World Natural (or Natural plus Cultural) Heritage', as provided for by Articles 3 and 2 of the 1972 *Convention concerning the Protection of the World Cultural and Natural Heritage* (UNTS 15511); (b) a 'Biosphere Reserve', as provided for by the UNESCO Man and the Biosphere (MAB) Program; (c) a 'Wetland of International Importance', as provided for or suggested by Article 2 of the 1971 *Convention on Wetlands of*

International Importance especially as Waterfowl Habitat (UNTS 14583); and (d) a 'Regional Sea', as provided for by the UNEP Regional Seas Program.

Article 6. We express our intention of considering the possibility of acting jointly in working toward any special-area designation described in Article 5.

Article 7. We express our intention to permanently demilitarize any protected areas we establish within our half of the DMZ and its contiguous NBZ or SBZ, doing so within the framework of Article 60 of the 1977 *Protocol [I] Additional to the 1949 Geneva Conventions and Relating to the Protection of Victims of International Armed Conflicts* (UNTS 17512).

Article 8. We express our intention to remove any known mine fields and other hazards from any protected areas we establish within our half of the DMZ and its contiguous NBZ or SBZ in an environmentally sensitive manner, doing so within the framework of Article 2 of the 1953 *Korean Armistice Agreement*; and also in the spirit of Article 1 of the 1997 *Convention on the Prohibition of the Use, Stockpiling, Production and Transfer of Anti-personnel Mines and their Destruction* (UNTS 35597). [A subsequent addition: And similarly of the 2008 Convention on Cluster Munitions (UNTS 47663).]

Article 9. We express our intention to have at least semi-annual consultative meetings on the progress of this **MoU** between the Ministers of the Environment of the DPRK and ROK or their respective designees, the venue of such meetings to be at a mutually acceptable neutral site, or else to alternate between Pyongyang and Seoul; and to which, at our joint discretion, observers may be invited to represent the *International Crane Foundation, DMZ Forum*, IUCN, UNEP, or other relevant intergovernmental agency or nongovernmental organization of our joint choosing.

Article 10. We express our intention that our long-term goal, in principle, is to consummate one or more formal agreements between the DPRK and ROK that would establish one or more transboundary protected areas for peace and nature in those instances where our unilaterally established contiguous protected areas coincide with a cross-border habitat that would most sensibly be managed as a joint endeavor. Such transboundary protected area(s) would thus be managed on a day-to-day basis by its own bilateral commission that enjoys a certain level of autonomy in its routine technical operations.

FINAL PROVISIONS

Article 11. There shall be two identical authentic copies of this **MoU** prepared in the Korean language, each to be signed by both states parties, one each for retention by the DPRK and ROK.

Article 12. The Annexes to this **MoU** form an integral part of this **MoU**.

Article 13. Once signed, this **MoU** shall remain valid until replaced by a formal bilateral treaty of comparable intent; or until either the DPRK or the ROK formally withdraws from it in writing.

SIGNED:

On behalf of the **Democratic People's Republic of Korea** (DPRK):

Signature: ____________________________

Name: ____________________________ Title: ______________________

At: ____________________ On: ____________

On behalf of the **Republic of Korea** (ROK):

Signature: ____________________________

Name: ____________________________ Title: ______________________

At: ____________________ On: ____________

At: ____________________ On: ____________

Appendix 7.5.1 Annex 1 Potential Sites Under Consideration

1.a. **The DPRK Lowland Protected Area in Kangwon Province**
2.a. **The ROK Lowland Protected Area in Gyeonggi Province**

These two contiguous largely low wetland areas under consideration for protection are situated ca 60 km northeast of Panmunjom. They are important as a migratory staging area or wintering ground for a number of migratory waterfowl following the Northeast Asian Flyway (thereby primarily involving the DPRK, ROK, China, and Russia). Indeed, these two areas are crucial to the survival of the red-crowned crane (*Grus japonensis*; IUCN Endangered) and the white-naped crane (*Grus vipio*; IUCN Vulnerable). Other threatened bird species that benefit from these two areas are the Chinese egret (*Egretta eulophotes*; IUCN Vulnerable), the black-faced spoonbill (*Platalea minor*; IUCN Endangered), and possibly also the hooded crane (*Grus monacha*; IUCN Vulnerable). Mammals threatened with extinction in these two areas include the Siberian musk deer (*Moschus moschiferus*; IUCN Vulnerable). The two areas support numerous species of indigenous fish, the latter providing a highly valuable source of fish where needed elsewhere in the DPRK and ROK for restocking in waters from which they have become extirpated. These two contiguous protected areas under consideration, which might each be 50,000 ha or more in size, have potential for both ecotourism and cultural tourism; and could perhaps fall within IUCN Category II or IV (cf. *Annex 2*) [Appendix 7.5.2].

2.a. **The DPRK Mountain Protected Area in Kumgansan Province**
2.b. **The ROK Mountain Protected Area in Gangwon Province**

These two contiguous largely temperate-forest upland areas under consideration for protection are situated ca 50 km southwest of the eastern terminus of the DMZ. They are important for the survival of a number of threatened mammalian species, including the Himalayan black bear (*Ursus thibetanus*; IUCN Vulnerable) and the Siberian musk deer (*Moschus moschiferus*; IUCN Vulnerable). Birds threatened with extinction that make use of these two areas include especially the red-crowned crane (*Grus japonensis*; IUCN Endangered). The area under consideration here by the DPRK already includes the Mount Kumgang National Park (60,000 ha; IUCN Category II), ca 30 km northwest of the Military Demarcation Line. The area under consideration here by the ROK already includes the Seoraksan National Park (39,800 ha; IUCN Category II), ca 40 km southeast of the Military Demarcation Line. Both of these existing protected areas are currently under consideration as World Heritage Sites. These two contiguous protected areas under consideration would in effect functionally connect those two existing national protected areas, thereby constituting a generally north-south mountainous ridge-line wildlife corridor especially beneficial to large mammals and other wildlife. These two contiguous protected areas under consideration, which might each add 20,000 ha or more to the two already existing protected areas, have potential for both eco-tourism and cultural tourism; and could perhaps fall within IUCN Category II or IV (cf. *Annex* 2) [Appendix 7.5.2].

Appendix 7.5.2 Annex 2 IUCN Protected Area Categories

All IUCN categories: Protected areas of land and/or sea dedicated to the protection and maintenance of biological diversity, and of natural and associated cultural resources.

IUCN Category Ia: 'Strict Nature Reserve', being a protected area managed mainly for science. This is an area of land and/or sea possessing some outstanding or representative ecosystems, geological or physiological features, and/or species, available primarily for scientific research and/or environmental monitoring.

IUCN Category Ib: 'Wilderness Area', being a protected area managed mainly for wilderness protection. This is a large area of unmodified or slightly modified land and/or sea, retaining its natural character and influence, without permanent or significant habitation, which is protected and managed so as to preserve its natural condition.

IUCN Category II: 'National Park', being a protected area managed mainly for ecosystem protection and recreation. This is a natural area of land and/or sea, designated to (1) protect the ecological integrity of one or more ecosystems for present and future generations, (2) exclude exploitation or occupation inimical to the purposes of designation of the area, and (3) provide a foundation for spiritual, scientific, educational, recreational, and visitor opportunities, all of which must be environmentally and culturally compatible.

IUCN Category III: 'Natural Monument', being a protected area managed mainly for conservation of specific natural features. This is an area containing one, or more, specific natural or natural/cultural feature which is of outstanding or unique value because of its inherent rarity, representative or aesthetic qualities, or cultural significance.

IUCN Category IV: 'Habitat/Species Management Area', being a protected area managed mainly for conservation through management intervention. This is an area of land and/or sea subject to active intervention for management purposes so as to ensure the maintenance of habitats and/or to meet the requirements of specific species.

IUCN Category V: 'Protected Landscape/Seascape', being a protected area managed mainly for landscape/seascape conservation and recreation. This is an area of land, with coast and sea as appropriate, where the interaction of people and nature over time has produced an area of distinct character with significant aesthetic, ecological, and/or cultural value, and often with high biological diversity. Safeguarding the integrity of this traditional interaction is vital to the protection, maintenance, and evolution of such an area.

IUCN Category VI: 'Managed Resource Protection Area', being a protected area managed mainly for the sustainable use of natural ecosystems. This is an area containing predominantly unmodified natural systems, managed to ensure long-term protection and maintenance of biological diversity, while providing at the same time a sustainable flow of natural products and services to meet community needs.

Note: For more detailed IUCN definitions and guidelines, cf.:
www.unep-wcmc.org/protected_areas/categories/index.html

Chapter 8
The Question of Globalization

8.1 Introduction

Examined here are current developments in globalization as these might influence developments in environmental security. I need not dwell at any length upon the dynamics of globalization itself since this has been well summarized by James Rosenau (1996); a useful brief historical perspective on the subject has recently been presented by Emma Rothschild (1999) and Paul Kennedy (1993) has explained at length how we should be preparing for a globalized twenty-first century. Neither is there a need for me to dwell at any length upon the concept of environmental security since I have done this a number of times elsewhere (Westing 1986, 1989b, 1991, 1999a).

Nonetheless, it will be valuable for me to begin by offering very brief synopses of the current dynamics of **globalization** on the one hand and the current dynamics of **environmental security** on the other. These two summaries readily lead us to five further issues: (1) whether the increasing tempo of globalization influences the increasing tempo in the deterioration of environmental security; (2) whether the positive aspects of globalization outweigh the negative ones; (3) whether the reversal of globalization would have a significant impact on environmental security; (4) whether globalization might be regulated so as to improve environmental security; and, finally (5) a suggested response to globalization.

This Chapter is reproduced from the author's Entry #347 provided in Chap. 2, with the original title, 'Globalization *vis-à-vis* environmental security'. It is used here by permission of the International Society of Naturalists, the copyright holder, as given on 15 March 2013. It had been an invited working paper, for 'The International Colloquium on Coming to Terms with Globalization: Possible Actions and Options for the Body Politic' of the Evangelische Akademie Loccum, Rehburg-Loccum, Germany, 10–12 December 1999. He is pleased to acknowledge suggestions from Carol E. Westing.

A. H. Westing, *From Environmental to Comprehensive Security*,
SpringerBriefs on Pioneers in Science and Practice 13,
DOI: 10.1007/978-3-319-00687-1_8, © The Author(s) 2013

8.2 The Dynamics of Globalization and of Environmental Security

8.2.1 Globalization

The ever quickening pace and growing diversity of globalization over the past several decades can be demonstrated by a number of objective (though certainly not thoroughly independent) measures. In the realm of **economics**, we are witnessing remarkable expansions in world trade, in interstate monetary transfers, in interstate capital investments, in multinational (transnational) corporations, and in organized transnational (metanational) crime. In the realm of **communication**, we are witnessing remarkable expansions both in one-way interstate contacts (via radio, television, and printed materials) and in two-way interstate contacts (via telephone, telefax, e-mail, and touristic or other person-to-person liaisons), all such contacts facilitated by the remarkable spread of English as the global *lingua franca*. In the realm of **knowledge and information**, we are witnessing remarkable expansions in the interstate exchange of scholars and experts, in the worldwide availability of scholarly and technical journals and reports, and in the global spread of culinary, musical, clothing, and other local customs (cf., e.g., Mooney 1998). And in the realm of **cooperation**, we are witnessing remarkable expansions in interstate governance (via multilateral humanitarian, arms control, human rights, and environmental treaties), in transnational collaboration (via environmental, human rights and other nongovernmental organizations [NGOs]), and in combined intergovernmental/nongovernmental actions.

8.2.2 Environmental Security

The ever quickening pace and growing diversity of deterioration in environmental security over the past several decades can also be demonstrated by a number of objective (though again not thoroughly independent) measures. In the realm of **natural-resource utilization**, we are witnessing remarkable worldwide expansions in the unsustainable exploitation of agricultural soils, tropical timber trees, rangeland grasses, ocean fisheries (both the higher and lower trophic levels), and freshwater supplies. In the realm of **waste-sink utilization**, we are witnessing remarkable worldwide expansions in the unsustainable disposal of gaseous, liquid, and solid wastes, prominent among them the unsustainable discharge into the atmosphere of carbon dioxide and other so-called greenhouse gases. In the realm of **biodiversity and related habitat protection**, we are witnessing remarkable worldwide habitat disruptions and losses in plant and animal species. And, finally, in the realm of **social security**—which is inextricably linked in both directions with environmental security—we are witnessing remarkable increases: (1) in human numbers; (2) in livestock numbers (such gains strongly linked with wildlife losses); (3) in intrastate (internal, non-international) armed conflicts (wars); (4) in refugee flows (both political and environmental, both intrastate and interstate); (5) in deadly communicable disease frequencies (e.g., in malaria, tuberculosis, yellow fever, and acquired immune deficiency syndrome [AIDS]); (6) in illiteracy; (7) in economically failing states (e.g., those 100 or more in which both GNP/*caput* and income/*caput* have been consistently declining for the past decade or more); (8) in the spread or intensification of unemployment, chronic hunger, and abject poverty; (9) in the ever widening gaps between the rich and the poor, both as between and within states; and (10) in a declining United Nations Development Programme (UNDP; New York) 'Human Development Index' for more than 30 states. Finally, scant comfort can be gained from the essentially unchanging prevalences in the very widely entrenched governmental autocracy and corruption.

8.3 The Influence of Globalization on Environmental Security

Several recent analyses shed some important light on the adverse impacts of globalization on environmental security, especially in relation to public health on the one hand, and to the global economic system on the other.

8.3.1 Public Health

Julio Frenk et al. (1997) and Anthony McMichael et al. (1999), among others, argue persuasively that, despite some important advances (e.g., in the spread of vaccinations, oral rehydration therapy, antibiotics, and medical information), globalization has in fact undermined the public health status of many states throughout the world. To provide some examples, infectious diseases are circulating more widely, and, as already noted, the worldwide incidences of such major killers as malaria, tuberculosis, yellow fever, and AIDS are rising substantially. The international trade in tobacco, as in illegal drugs, is expanding remarkably, leading to the spread of ever more serious associated public health problems. Multinational corporations are moving their polluting and otherwise dangerous production facilities to states with lax public health and environmental protection standards. Disease microörganisms throughout the world are developing resistances to antibiotics. Chronic malnutrition is on the increase in many Third-world states. The fast-growing Third-world urban populations are increasingly exposed to unsanitary (polluted) conditions. And, finally, the spread of westernized diets is beginning to prove widely detrimental (as evidenced, e.g., by increases in cardiovascular diseases, diabetes, and breast cancers).

8.3.2 Economics

Richard Falk (1995) and Clem Tisdell (1999), among others, make it amply clear that economic globalization in its various ramifications is largely detrimental to environmental security. To begin with, the accepted wisdom that continued economic growth is needed for the creation of employment opportunities and other benefits to society undermines the very necessary efforts to achieve an environmentally sustainable steady-state economy. Indeed, the global natural-resource base—both its renewable resources and its renewable sink capacities—simply does not suffice to accommodate any widespread improvements in living standards even approaching those now enjoyed in the 27 or so highly industrialized states.

Especially antithetical to the requirements of environmental security is the spread of freer markets, which makes it difficult, if not virtually impossible, for member states to restrict environmentally deleterious interstate trade—with the freer markets so forcefully encouraged by the 1947 General Agreement on Tariffs and Trade (GATT; Geneva), its successor, the 1995 World Trade Organization (WTO; Geneva), and the 1994 North American Free Trade Agreement (NAFTA; Washington). The widespread acceptance by Third-world states of the so-called structural adjustment conditions imposed by the International Bank for Reconstruction and Development (World Bank; Washington) and the International Monetary Fund (IMF; Washington) further undermines environmental security, not only by their emphasis on freer trade, but also often by encouraging conversions from small-scale subsistence farming to large-scale commercial agriculture for the production of exportable commodities. Similarly, it has been recognized that environmental security suffers in states being newly brought into the international economic system, in large part owing to the continuously strong emphasis on economic growth. This was seen to be the case especially in the early stages of such integration, but may well carry on into the medium-term and even long-term futures, particularly as the natural-resource-based carrying capacity of a state is in the process likely to be exceeded.

8.4 Positive *versus* Negative Aspects of Globalization

It seems clear that environmental security in the aggregate is undermined by economic and other aspects of globalization. But all is not bleak on the globalization front. Thus, we certainly must not overlook the spread of knowledge and information among the states and the peoples of the world. Such spread permits—and should presumably thus be conducive to—more environmentally enlightened practices at local, national, regional, and worldwide levels. Then there is the proliferation and increased networking of both national and international NGOs, especially in the fields of environment and of human rights. The voice that such issue-oriented NGOs now give to people provides an increasingly important influence on (counterbalance to) national governments and their intergovernmental agencies. The emerging collaborations between nongovernmental organizations and governments are another welcome phenomenon (Reinicke 1999–2000). And the formation of a 'Global Peoples' Assembly' has recently been proposed for the special purpose of democratizing the international order, that is, the political dimensions of globalization (Falk and Strauss 1999).

8.5 A World Without Globalization

Is globalization at the root of all the world's current ills, as seems to be suggested by such advocacy groups as the International Forum on Globalization (IFG; San Francisco; 1994–) or the Association for the Taxation of Financial Transactions for the Benefit of Citizens (ATTAC; Paris; 1998–)? At least with respect to environmental security, it is abundantly clear that certain aspects of globalization are detrimental to environmental security whereas other aspects of it are supportive of environmental security. But I would venture to suggest that even if the process of globalization could be reversed—indeed, even if it could be largely eliminated—that would not address the fundamental challenges to the achievement of environmental security.

At the very heart of the growing deterioration in our environmental security are the seemingly ever expanding human and associated livestock numbers. Inexorable concomitants of these linked expansions are the expanding arrogations of what remains of the natural world, of the unsustainable exploitation of the world's major renewable natural resources, and of the unsustainable exploitation of the world's major waste depositories (sinks).

At another level, groundbreaking analyses by Jared Diamond (1997) and David Landes (1999) make it agonizingly clear that some combination of biogeographical (environmental), cultural, and historical differences between the rich states and the poor ones have been largely responsible for their starkly contrasting levels of development. These structurally unavoidable and deeply imbedded differences between the present haves and have-nots will make it extraordinarily difficult, if not virtually impossible, to equalize them.

At yet another level, the growing deterioration in our environmental security is so difficult to address owing to certain enduring human traits, especially: (1) our widespread inability to submit to nonviolent systems of law and justice, either at the national or interstate levels; (2) our widespread submission to totalitarian and corrupt governments, both local and national; and (3) our continuing recourse to armed conflict with deadly and destructive intent for the ultimate settlement of our many disputes, both sub-national and national.

The various factors antithetical to environmental security just outlined—factors that are at the very core of the poor state and continued deterioration in our environmental security—all predate the great expansion in globalization of recent years, and would presumably all remain functional even if the various processes of globalization could by some legerdemain be made to vanish.

8.6 The Regulation of Environmental Security and of Globalization

Globalization is, of course, not occurring in a vacuum. Processes that might be termed nationalization, localization, internationalization, and regionalization are also in progress with respect to the state system. Thus, **nationalization** has been occurring with enormous vigor in recent decades and most often with zealous attachment to national sovereignty. In fact, the number of sovereign states merely since 1950 has progressed from about 88 to the present 192 or so. **Localization**, often a prelude to the formation of new states, but to some extent a process of its own, has also been flourishing in recent decades. One poignant example of the burgeoning of localization is the growing numbers of intrastate armed conflicts. **Internationalization**, the collaborating by states via universal multilateral treaties, and often the concomitant formation of intergovernmental agencies, has also been flourishing in recent decades (examples already noted earlier being WTO and NAFTA). But it must be pointed out that such internationalization has not been accompanied by any substantial relinquishment of national sovereignty—witness, for example, the continuing great paucity of states (a mere two dozen or so) that have submitted to the International Court of Justice [World Court; the Hague] on a compulsory and unconditional basis; or the reluctance of the necessary 60 states to permit the 1998 Statute of the International Criminal Court to become a reality, only 5 as of December 1999.[1] And **regionalization**, a geographically limited subcategory of internationalization, has also been occurring to a more or less modest extent, especially in the field of environment (more on this below).

There is considerable room for the enactment and/or enforcement of regulations and associated actions and programs at the state level that would be directly supportive of environmental security. Despite the unlikelihood of such corrective actions by states—according to Falk (1995), owing to some combination of lack of will and lack of capacity for the public sector to become more intrusive—it is nonetheless important to enumerate at least some of the most urgent needs. Thus, all states must develop or expand programs which are supportive of family planning and population stabilization (or even of reduction, as necessary), both domestically and beyond, the latter via appropriate intergovernmental agencies (e.g., the United Nations Population Fund [UNFPA; New York]). Important for all states, but especially so for the advanced industrial ones, environmental protection legislation enacted by them must cover not only domestic actions by individuals and corporations, but also all those carried out by its nationals and corporations in

[1] The 1998 Statute of the International Criminal Court came into force in 1999 (UNTS 38544).

other states or, it must be stressed, in the extra-territorial portions of the global biosphere. Any national environmental protection constraints should be applicable not only to goods destined for export, but also to imported goods. And it would be a most instructive and useful exercise for all states to annually publish a national environmental ('green') accounting, with emphasis on indicators of progress (if any) toward sustainability, as was pioneered by the Netherlands in 1985.

It is crucial that the member states of intergovernmental agencies such as the World Bank, IMF, and WTO act collectively, not only to make their deliberations more transparent, but especially to 'green' them up more rigorously, so that environmental considerations will no longer be—as they are now—subservient to free-trade considerations (cf., e.g., French 1999). Comparable 'bottom-line' environmental considerations should also be fundamental to the actions of such restricted multilateral groupings as the European Free Trade Association (EFTA; Geneva; 1960–), the European Union (EU; Brussels; as established by the 1992 Maastricht Treaty; UNTS 30615), and the Organisation for Economic Co-operation and Development (OECD; Paris; 1961–).

Multinational corporations are not so readily amenable to regulation throughout their often far-flung spheres of operation as one might wish, but NGOs can apply at least some pressure on them by publicizing their environmental actions or human rights records, both good and bad. Greenpeace (Amsterdam; 1970–) has become well known for vigorously publicizing what it considers to be grievous corporate environmental transgressions. Other less flamboyant NGOs have also begun to interest themselves in various segments of industry. For example, Eco-tex (Köln; 1992–) monitors the textile industry (production, manufacture, and trade) from an environmental standpoint and offers 'green' certification; the Rainforest Alliance (New York; 1987–) provides 'green' certification for sustainably harvested tropical timbers; and the Fair Labor Association (Washington; 1999–) is in the process of publicizing the human rights records of the major multinational clothing and footwear manufacturers. Apropos such labeling, the Canadian government has since 1988 been issuing an 'EcoLogo' to products that pass muster. Finally it must be added here that industry itself has also begun to become more environmentally responsible (Nash and Ehrenfeld 1996; Spencer-Cooke 1999; White 1999). Manifestations of this include greater corporate environmental transparency and the adoption of environmental codes of conduct via such corporate coalitions as the International Chamber of Commerce (Paris; 1972–) and the World Business Council for Sustainable Development (Geneva; 1990–).

As to foreign aid, it would certainly be gracious for all states (although forgiving the several dozen thoroughly bankrupt ones) to annually donate some modest fraction of their GNP (0.7 % is often suggested) for foreign development assistance. None (or very little) of such aid should be of a bilateral nature, but should rather be channeled through such intergovernmental agencies as UNDP, the United Nations Conference on Trade and Development (UNCTAD; Geneva), or especially the Global Environment Facility (GEF; Washington), with emphasis on capacity building. All such development assistance should without fail have the attached conditionality of a workable system of strict fiscal responsibility and environmental accountability. And finally, states could usefully join with their neighbors to form regional compacts, as dwelt upon below.

8.7 Regionalization as an Approach to Environmental Security

To me the most productive approach to strengthening environmental security in the face of globalization is via widespread regionalization, based on regions (often sub-regions in diplomatic parlance) that have some logical coherence both geographically and environmentally—what I have termed 'ecogeographical' regions (Westing 1989a). More specifically, environmental security should for each state really begin at home (Westing 1996, 1999b), although by necessity almost always together with the ecogeographical region of which it is an integral and unavoidable part (Byers 1991; VanDeveer and Dabelko 1999; Westing 1989a, 1994). Such an ecogeographical security regime will encompass a domain which the included governments and people can readily grasp, with which the regional inhabitants can presumably identify, and which has a logical watershed (catchment basin) or other environmental, and often also social or cultural, coherence. Thus, social security built upon environmental security—that is to say, sustainable development within a framework of social justice—not only must, but can most readily, be achieved within the national context of each state in concert with its immediate ecogeographical partners. National security in the traditional sense—protection from any external enemies—has to be retained of course, but only within the broader framework of sustained protection from hunger, maltreatment, internal strife, and despair. And, it seems clear, such aims have the greatest opportunity for realization at the national and ecogeographical levels. The Baltic Sea region (VanDeveer and Dabelko 1999; Westing 1989a, b), the Horn of Africa (Westing 1999b), the Red Sea region (Westing 1999b), the Aral Sea region (Westing 1994), the Tigris-Euphrates region (Westing 1994), and the Korean peninsula (Westing 1998–1999) provide a few specific examples.

That is not to suggest that there are no environmental security issues of global scope. Thus by way of example, to achieve sustainable and equitable exploitation of such common-domain natural resources as the atmosphere or the high seas and its seabed, or to rid the world of nuclear, biological, or chemical weapons or of land mines and other explosive remnants of war, will require broad multilateral action. Past experience suggests that to attain globally sustainable solutions will not be easy, and to make them equitable on top of that virtually impossible. But as more ecogeographical regions achieve their own environmental security, the involved states will be able to enter multilateral negotiations on global issues from positions of greater strength and confidence.

Many of the awesome challenges to environmental security—over-population (i.e., human numbers in excess of available natural resources and sink capacities as well as of available employment opportunities), the many intrastate armed conflicts with their huge numbers of casualties and associated social and environmental disruptions, and ever-growing numbers of internal and cross-border refugees, whether political or environmental, to name but a few of the more obvious ones—are most efficiently addressed at national and ecogeographical levels.

8.8 Conclusion

Globalization—especially in its multifarious economic and communication dimensions—has been accelerating with a vengeance. Nonetheless **nationalization**—as evidenced by the increasing numbers of states and their tenacious grips on national sovereignty—seems as entrenched as ever. For its part, **environmental security** is seriously deteriorating on a worldwide basis. Such deterioration ultimately derives from human numbers having by now far outstripped the carrying capacity of the global biosphere, this in concert with a high proportion of inept, indifferent, autocratic, and corrupt national regimes. Thus, the deleterious factors leading to a dissipation of environmental security are largely independent of globalization, but are, as we have seen, reinforced by it. And as I try to make a case for in the text above—in addition to the 'greening' of world trade through environmentally sensitive governmental and intergovernmental regulations—the fundamental approach to ameliorating this dire state of affairs is not so much, in my view, to 'think globally, act locally', but rather to '**think nationally, act regionally**'. Finally, the approach to global sanity being called for here would certainly be facilitated by the widespread nurturing of an ethic (some would suggest a spiritually based ethic) that recognizes not only human rights, but also the rights of nature.

References

Byers, B. 1991. Ecoregions, state sovereignty and conflict. *Bulletin of Peace Proposals* [now *Security Dialogue*] (Oslo) 22(1):65–76.

Diamond, J. 1997. *Guns, Germs, and Steel: the Fates of Human societies*. New York: W.W. Norton, 480 pp.

Falk, R. 1995. Environmental protection in an era of globalization. *Yearbook of International Environmental Law* (Oxford) 6:3–25.

Falk, R. & Strauss, A. 1999. Globalization needs a dose of democracy. *International Herald Tribune* (Paris) 1999(36263):8.

French, H. 1999. Trade ministers must not forget or neglect the environment. *International Herald Tribune* (Paris) 1999(36308):8.

Frenk, J. et al. 1997. The future of world health: the new world order and international health. *British Journal of Medicine* (London) 314(7091):1404–1407.

Kennedy, P. 1993. *Preparing for the Twenty-first Century*. New York: Random House, 429 pp.
Landes, D.S. 1999. *Wealth and Poverty of Nations: Why Some are so Rich and Some so Poor*. rev. edn. New York: W.W. Norton, 658 pp.
McMichael, A.J. et al. 1999. Globalization and the sustainability of human health: an ecological perspective. *BioScience* (Reston, VA) 49(3):205–210.
Mooney, H.A. 1998. *The Globalization of Ecological Thought*. Oldendorf/Luhe, Germany: Ecology Institute, Excellence in Ecology Series No. 5, 153 pp.
Nash, J. & Ehrenfeld, J. 1996. Code green: business adopts voluntary environmental standards. *Environment* (Philadelphia, PA) 38(1):16–20,36–45.
Reinicke, W.H. 1999–2000. The other world wide web: global public policy networks. *Foreign Policy* (Washington) 1999–2000(117):44–57.
Rosenau, J.N. 1996. Dynamics of globalization: toward an operational formulation. *Security Dialogue* (Oslo) 27(3):247–262.
Spencer-Cooke, A. 1999. Treading lightly. *People & the Planet* (London) 8(3):6–9.
Rothschild, E. 1999. Globalization and the return of history. *Foreign Policy* (Washington) 1999(115):106–116.
Tisdell, C. 1999. *Biodiversity, Conservation and Sustainable Development: Principles and Practices with Asian examples*. Cheltenham, UK: Edward Elgar, 263 pp.
VanDeveer, S.D. & Dabelko, G.D. 1999. Redefining security around the Baltic: environmental issues in regional context. *Global Governance* (Wellesley, MA) 5(2):221–249.
Westing, A.H. 1986. Expanded concept of international security. In: Westing, A.H. (ed.). *Global Resources and International Conflict: Environmental Factors in Strategic Policy and Action*. Oxford: Oxford University Press, 280 pp: pp 183–200.
Westing, A.H. 1989a. Environmental approaches to regional security. In: Westing, A.H. (ed.). *Comprehensive Security for the Baltic: an Environmental Approach*. London: Sage Publications, 148 pp: pp 1–14.
Westing, A.H. 1989b. Regional security in a wider context. In: Westing, A.H. (ed.). *Comprehensive Security for the Baltic: an Environmental Approach*. London: Sage Publications, 148 pp: pp 113–121.
Westing, A.H. 1991. We have to seek security in its broadest sense. *World Health Forum* (World Health Organization, Geneva) 12(2):137–139.
Westing, A.H. 1994. Environmental change and the international system: an overview. In: Calließ, J. (ed.). *Treiben Umweltprobleme in Gewaltkonflikte?: Ökologische Konflikte im internationalen System und Wege zur Kooperation*. Rehburg-Loccum, Germany: Evangelische Akademie Loccum, Loccumer Protokolle, No. 21/94, 354 pp: pp 207–228.
Westing, A.H. 1996. Canadian security in a broadened context. *Canadian Defence Quarterly* (Toronto) 26(2):13–22.
Westing, A.H. 1998–1999. A transfrontier reserve for peace and nature on the Korean peninsula. *International Environmental Affairs* (Hanover, NH) 10(1):8–17.
Westing, A.H. 1999a. Biodiversity loss and its implications for security and armed conflict. In: Cracraft, J. & Grifo, F.T. (eds). *The Living Planet in Crisis: Biodiversity Science and Policy*. New York: Columbia University Press, 311 pp: pp 209–216.
Westing, A.H. 1999b. Conflict versus cooperation in a regional setting: lessons from Eritrea. In: Suliman, M. (ed.). *Ecology, Politics and Violent Conflict*. London: Zed Books, 298 pp: pp 273–285.
White, A.L. 1999. Sustainability and the accountable corporation: society's rising expectations of business. *Environment* (Philadelphia, PA) 41(8):30–43.

Chapter 9
Environmental Refugees: A Stark Reminder

It is noted that the number of more or less permanently displaced persons throughout the world (now of the order of 1 % of the total human population) continues to increase at a rate of approximately 3 million per year; the situation in Africa is especially grave, with the number of displaced persons there (now of the order of 3 % of the African population), continuing to increase at a rate of approximately 1.5 million per year. Human displacement—which can be seen to originate largely in rural areas—results primarily from one or more of three factors, namely escape from persecution, escape from military activities, or escape from inadequate means of subsistence. A number of examples from Africa are provided of the social and political consequences of human displacement, with emphasis on conflict situations at the sites of relocation.

It is further noted that the numbers of displaced persons continue to grow relentlessly despite there being no discernible rise in persecution or in military activities, and despite the long-sustained ameliorative efforts and financial assistance by intergovernmental agencies and others.

It is accordingly suggested that the major cause of the continuing increase in the numbers of displaced persons is an ever-growing imbalance between population numbers and the human carrying capacity of the land. Population increases lead to smaller per caput natural resource bases, a predicament exacerbated by over-use—and thus degradation—of the land and its natural resources. In the arid and semi-arid regions of Africa, over-use of the land most often takes the form of overgrazing, leading to land degradation that is severe enough to be referred to as desertification. It is concluded that to achieve sustainable utilization of the land

This Chapter is reproduced from the author's Entry #257 provided in Chap. 2, with the original title, 'Population, desertification, and migration'. It is used here by permission of the Foundation for Environmental Conservation, the copyright holder, as given on 13 March 2013 (and verified on 18 March 2013). It is based on an invited paper presented to 'The International Symposium on Desertification and Migration', Almeria, Spain, 9–11 February 1994, of the United Nations Intergovernmental Negotiating Committee for a Convention to Combat Desertification [INCD]. The Author is most pleased to acknowledge useful suggestions from Johannes B. Opschoor and Carol E. Westing.

A. H. Westing, *From Environmental to Comprehensive Security*, SpringerBriefs on Pioneers in Science and Practice 13,
DOI: 10.1007/978-3-319-00687-1_9, © The Author(s) 2013

and its natural resources will necessitate the integrated attainment of environmental security and societal security—the latter inter alia requiring participatory governance, non-violent means of conflict resolution, and especially population controls.

9.1 Background

The number of displaced persons throughout the world—at any one time, now approximately 1 % of the total human population—continues to increase by perhaps 3 million per year (Westing 1992, Table I). The situation is especially acute in Africa, where the current level of human displacement is of the order of 3 %, and the rate of increase is roughly 1 million per year (Westing 1992, Table II). Almost half of all displaced persons—whether refugees (asylum seekers), migrants (opportunity seekers), or some combination thereof—cross a national border, that is, leave their country of origin (Westing 1992, Table I).

Persons become more or less permanently displaced from their areas of habitual residence for various reasons, among them especially: (a) to escape persecution for reasons of race, religion, nationality, their membership in a particular social group, or their political opinion; (b) to escape the dangers of armed conflict, including massacres and other severe internal upheavals; and (c) to escape inadequate means of subsistence.

As for the three most prominent causes of the longterm population displacement just noted, the incidence of persecution (political oppression and associated abuses of human and civil rights) is extraordinarily widespread, although its frequency seems not to have been increasing in recent years (Table 9.1). Similarly, the incidence of armed conflict is extraordinarily widespread, although its frequency also seems not to have been increasing in recent years (Table 9.2); or even in recent decades (Westing 1982). By contrast, the means of subsistence—already marginal for large numbers of people throughout the world—continue to deteriorate in an increasing number of countries (Westing 1990). Human population numbers continue to soar in the non-industrialized world, inevitably accompanied by declining *per caput* natural-resource bases; and often (especially in Africa) together with declining national *per caput* productivities.

The assumption can thus be made that the ever-increasing numbers of persons in the non-industrialized world who are migrating to their country's cities or even to other countries—often to countries in the industrialized world—are in large part doing so to escape from inadequate means of livelihood (such migration being fueled to some extent by a growing awareness of existing opportunities elsewhere). Population increases in rural areas soon result in human numbers that cannot be locally sustained—a situation which is usually exacerbated by the utilization of the available renewable resources (water, soil, forage, wood, fish, and wildlife) at damaging and diminishing, i.e., non-sustainable, rates. Water scarcity (resulting from the over use of water) is rapidly becoming a particularly serious problem in

Table 9.1 Political oppression and despotism in the world[1]

Years	Oppressed population worldwide (%)	Despotic regimes worldwide (Number)	Despotic regimes in Africa (Number)
1980	42	57	29
1981	43	59	29
1982	44	62	29
1983	44	64	31
1984	41	60	30
1985	42	56	32
1986	40	55	33
1987	39	53	32
1988	38	51	31
1989	41	68	39
1990	39	62	34
1991	33	51	30
1992	32	42	21
1993	31	39	18

Notes

(a) Percentage of oppressed population worldwide is based on an analysis of ca 168 states, derived from a composite score of political rights and civil liberties, that fall into the category 'not free' (for 1980, Gastil 1982: 4; for 1981–1992, McColm 1992a: 48; for 1993, McColm 1993: 4)

(b) Number of despotic regimes worldwide is based on an analysis of ca 168 states, and in Africa on ca 50 states; derived from a composite score of political rights and civil liberties, that fall into the category 'not free' (for 1980–1991, McColm 1991: 462–465 (as corrected); for 1992, McColm 1992b: 578; for 1993, McColm 1993: 625). It must be noted that more stringent criteria than in other years were employed for the 1989 values (Freedom House, New York, pers. comm., 27 September 1993)

various parts of the world, with northern Africa providing an especially acute example (Falkenmark 1989).

When once an imbalance between population numbers and the carrying capacity of the land develops, the persons thereby displaced are apt to be referred to as 'environmental refugees' or 'environmental migrants'.

9.2 Desertification

Countries that are characterized to a substantial extent by very low and sporadic rainfall—and accordingly by sparse vegetation—have traditionally supported a low density of people in large portions of their rural areas. In Africa, for example, these people are often nomadic pastoralists, whose traditional means of subsistence have long taken into account the regionally meager and irregular availability of water. Here the utilization of the land beyond sustainability leads to land degradation and, ultimately, desertification. Indeed, of the almost 500 million ha of land that has been degraded by human action throughout Africa (Table 9.3), about half has resulted primarily from livestock overgrazing (Table 9.3, Note b).

Table 9.2 Militarism in the world

Years	Armed conflicts worldwide (Number)	Armed conflicts in Africa (Number)	Military spending by the non-industrialized world (10^9 US$)	Military spending by the African states (10^9 US$)
1980	?	?	190	18
1981	?	?	200	18
1982	?	?	220	18
1983	?	?	220	19
1984	?	?	220	19
1985	?	?	220	17
1986	36	11	210	18
1987	36	11	210	17
1988	38	11	200	17
1989	32	9	200	18
1990	31	10	?	18
1991	30	11	?	17
1992	30	8	?	?

Notes

(a) Number of armed conflicts worldwide, and in Africa, is a count of all armed conflicts (both international and internal), that have resulted in a total of at least 1,000 battle-related fatalities (for 1986, Goose 1987; for 1987, Wilson and Wallensteen 1988; for 1988, Lindgren et al. 1989; for 1989–1992, Amer et al. 1993 [for 1989–1992, also Wallensteen and Axell 1993, some of whose values differ inconsequentially])

(b) Military spending by the non-industrialized world is based on ca 116 'developing' states; military spending by the African states is based on ca 46 states; the values are expressed in 1993 US dollars (ACDA 1991, Table I, with 1989 US$ $\times$ 1.165 = 1993 US$); military spending values by the African states for 1990–1991 are from the US Arms Control and Disarmament Agency (Washington, pers. comm., 2 February 1994)

Land degraded by human over-exploitation can result not only from population increases among the nomadic pastoralists and their herds, but also from diminution of the area of available rangelands. This diminution becomes the case because nationwide population pressures frequently lead to government policies that discourage nomadism and encourage an increased conversion of the rangelands to agriculture. Additionally, livestock policies may involve reserving large areas for modem ranching by wealthy cattle owners, almost inevitably at the expense of grazing options that had been open to the traditional pastoralists. Still other government actions might lead to conversion of lands that were previously devoted to subsistence agriculture by traditional means, to the mechanized production of cash crops for export.

The various encroachments just enumerated are especially short-sighted actions because those areas that are best suited to traditional rangeland utilization or

[1] The values in Table 9.1 appear to have changed little between 1993 and 2013 (http://www.freedomhouse.org/report/freedom-world-aggregate-and-subcategory-scores).

Table 9.3 Anthropogenic land degradation in Africa

Level of degradation	Land Area (10^6 ha)	Land Area (%)
Land degraded insignificantly (*total*)	2,472	83
Non-vegetated land	732	25
Vegetated land without agri- or silviculture	435	15
Vegetated land with agri- or silviculture	1,305	44
Land degraded significantly (*total*)	494	17
Vegetated land degraded lightly	174	6
Vegetated land degraded moderately	192	6
Vegetated land degraded strongly	124	4
Vegetated land degraded extremely	5	0.2
Total land area	2,966	100

Notes

(a) The levels of significant anthropogenic degradation of vegetated land are designated as: <1> 'light', which can be restored given minor changes in land-use practices; <2> 'moderate', which can be restored given major changes in land-use practices; <3> 'strong', which can be restored only at high cost; and <4> 'extreme', which cannot be restored

(b) The total of 494 million ha of vegetated land in Africa that has been significantly degraded by human action is primarily the result of: <1> overgrazing (49 %); ⟨2⟩ improper agricultural practices (25 %); <3> deforestation and other removal of vegetation (14 %); and <4> over-harvesting of vegetation for fuel and other domestic uses (13 %)

(c) The proximate mechanisms of degradation of the vegetated land in Africa are in essence: <1> water erosion (46 %); <2> wind erosion (38 %); <3> nutrient loss (9 %); <4> soil compaction (4 %); and <5> salinization (3 %)

(d) The presented values are derived from the data of UNEP/ISRIC (1990), in part as summarized by WRI (1992, Tables 19.3, 19.4)

traditional small-scale agriculture are apt to deteriorate under intensive large–scale exploitation (Table 9.3, Note b)—an additional generator of displaced persons. The site damage itself results primarily from soil erosion (by water or wind), and secondarily from 'nutrient dumping' (loss of soil nutrients in solution), soil compaction, and soil salinization (Table 9.3, Note c).

9.3 Social and Political Consequences of Environmental Migration

The social and political impacts of long-term environmental migration can be recognized and distinguished at a number of loci: (a) at the site of origin of the displaced persons—by the residual population; (b) at rural sites of destination within the nation—between the new arrivals and pre-established populations; (c) in the cities within the nation—among the new arrivals themselves or between them and pre-established residents; (d) in the non-industrialized foreign countries—between the new arrivals and the authorities or between the new arrivals and the indigenous residents; and (e) in the industrialized foreign countries—again, between the new arrivals and the authorities or between the new arrivals and the indigenous residents. Each of these five categories merits brief individual attention. (The environmental impact of refugee movements at the site of destination is reviewed in detail by Jacobsen 1994.)

9.3.1 Site of Origin

The social conditions in the rural areas of origin of environmental migrants must be assumed to have been sufficiently desperate to induce at least a partial exodus. The residual population is thus often left to depend upon a depleted and degraded resource base. Moreover, the residual population often consists of disproportionate numbers of destitute women, children, old people, and incapacitated individuals. In the event that an area which had previously been devoted to pastoralism is converted to agriculture, the displaced pastoralists might respond through armed rebellion, for instance, as occurred in the Sudan in the 1980s (Salih 1993: 125–126).

9.3.2 Domestic Rural Sites of Destination

When nomadic pastoralists abandon their degraded traditional areas to encroach on the traditional areas of others, friction arises and outright clashes are apt to occur. By way of example, increased violence of this sort between two Somali clans in the mid-1970s has been described (Molvær 1991: 177). Second, when small-scale traditional farmers and pastoralists are driven from their traditional lands by large-scale government-instituted or government-condoned schemes of mechanized agriculture, civil war can be the outcome, as appears to have been the case in the Sudanese Civil War of 1983— (Suliman 1992: 16–17, 22–23).

9.3.3 Domestic Urban Sites of Destination

The cities to which a large proportion of the migrants tend to gravitate (usually the nations' capital cities) cannot properly absorb the bulk of them. Sufficient job opportunities do not exist, nor do adequate housing, water and sewage services, hospitals, schools, and so forth. Huge slum areas or shanty towns have accordingly developed. The cities of northern Africa are among those of the non-industrialized world that are growing at staggering rates (Table 9.4). In some instances, the disenchanted urban squatters become a politically restive and even destabilizing force—as occurred, for example, in Sudan in the 1980s, especially in Khartoum and Port Sudan (Molvær 1991: 179–181).

Table 9.4 Northern African nations: population growth rates (total and urban)

State	Total population		Urban population	
	(%/yr)	(DT yrs)	(%/yr)	(DT yrs)
Algeria	3.0	23	4.8	14
Chad	2.4	29	6.3	11
Egypt	2.5	28	3.2	22
Ethiopia	3.1	22	5.3	13
Libya	4.1	17	6.3	11
Mali	2.6	27	3.8	18
Mauritania	2.4	29	7.3	9
Morocco	2.6	27	4.3	16
Niger	3.3	21	7.4	9
Somalia	3.1	22	5.6	12
Sudan	2.7	26	3.9	18
Tunisia	2.4	29	2.8	25
[*22 industrialized states* (*cf. Note e*)]	*0.6*	*116*	*0.8*	*87*

Notes

(a) Total population growth is for 1980–1991 (IBRD 1993, Annex Table 26), except for Libya and Somalia, which is for 1980–1990 (IBRD 1992, Annex Table 26)

(b) Urban population growth is for 1980–1991 (IBRD 1993, Annex Table 31), except for Libya and Somalia, which is for 1980–1990 (IBRD 1992, Annex Table 31)

(c) DT yrs = doubling time in years (where DT = $\log_n 2 \times 100 \div \%$ growth rate)

(d) Comparable data are unavailable for the remaining northern African nations (Djibouti, Eritrea, and Western Sahara). Algeria, Libya, Mauritania, Morocco, and Tunisia, comprise the Arab Maghreb [Western] Union

(e) '22 industrialized' states (provided here for purposes of comparison) is the mean of 22 'high-income economies' for 1980–1991 (IBRD 1993, Annex Tables 26 & 31)

9.3.4 Foreign Non-industrialized Sites of Destination

The foreign countries to which many of the displaced persons are migrating—in the case of northern African migrants, many to other (often more southerly) African countries and some to industrialized countries in Europe (cf. below)—are being subjected to increasing levels of migrant-induced economic, cultural, and political strains (Loescher 1992; Widgren 1990). For example, the enmity that brought about the Ethiopian–Somali War of 1977–1978 was greatly aggravated by border crossings from Somalia to Ethiopia by nomadic pastoralists (Molvær 1991: 177; Myers 1987: 18). In another instance, as a consequence of desertification, thousands of nomadic pastoralists from Mauritania migrated south into Senegal during the 1980s. This migration was abetted by Mauritania and, in 1989, resulted not only in armed border clashes, but also in deadly civil (ethnic) riots in both countries (Suhrke 1993: 30–31).

9.3.5 Foreign Industrialized Sites of Destination

With ever-growing numbers of migrants entering Europe, both legally and illegally, xenophobic manifestations are beginning to surface with increasing frequency, immigration restrictions are becoming more and more stringent, and attempts at addressing the perceived threats to security and stability at a deeper level are becoming more widespread (Golini et al. 1991; Hamilton and Holder 1991; Loescher 1992; Myers 1993; Ruthström 1990, 1991; Widgren 1989; Widgren 1990). For example, the growing problems associated with south-to-north migration across the Mediterranean Sea have recently led France, Italy, Portugal, and Spain to enter into a 'Four-plus-Five' consultative arrangement with Algeria, Libya, Mauritania, Morocco, and Tunisia; this has the aim of providing greater northern African political and economic stability and, thereby, of reducing the flow of migrants (Loescher 1992: 20). Despite such efforts—and in the face of shrinking labor markets (Salt 1993: 20–22)—the burgeoning extent of migration from the southern shores of the Mediterranean Sea to its northern ones (especially to France, Italy, Portugal, and Spain; also to Germany and Greece) has even led recently to the suggestion that, for better or worse, the countries of northern Africa are rapidly becoming a functional part of Europe (Fox 1993: 551–552).

9.4 Recommendations

The problem of long-term migrants—both internal and cross-border—keeps growing, despite there being no discernible rise in the threat of persecution or in military activity, despite the long-sustained ameliorative efforts on the part of intergovernmental agencies and nongovernmental organizations, and despite the considerable infusions of multilateral and bilateral financial aid to the migration-generating countries. It thus seems clear that the overall lack of success of these sundry efforts results from the fact that they do not address the root causes of the problem—to wit, the enduring pervasiveness of repressive forms of governance and the continuing resort to armed conflict, but, most especially, the ever-increasing population pressures and associated environmental strains.

In order to attain and sustain environmental security—sustainable utilization of resources, sustainable discard of wastes, and adequate protection of biodiversity—it does not suffice to deal with attempts at agricultural and rangeland improvement. Environmental security can only be attained and sustained within the framework of comprehensive human security, of which the former is an inextricable component (Westing 1989, 1991). Population growth must be curtailed—and, in many instances, even reversed—so that the long-term carrying capacity of the land is not exceeded. Equitable and otherwise non-corrupt participatory governance must be achieved at both local and national levels. Education, health care, and old-age security—now generally supported at unconscionably low levels (Table 9.5)—

Table 9.5 Northern African nations: some annual per caput public expenditures

State	Education (US$/capita)	Health (US$/capita)	Military (US$/capita)
Algeria	291	45	58
Chad	5	1	13
Egypt	51	9	44
Ethiopia	7	2	21
Libya	535	201	482
Mali	8	3	10
Mauritania	29	12	23
Morocco	57	10	47
Niger	13	8	7
Somalia	1	0.4	2
Sudan	31	2	31
Tunisia	101	36	34
[*28 industrialized states (cf. Note c)*]	*849*	*856*	*608*

Notes
(a) The values are for 1990, expressed in 1993 US dollars (Sivard 1993, Table II, with 1987 US$ × 1.272 = 1993 US$)
(b) Comparable data are unavailable for the remaining northern African nations (Djibouti, Eritrea, and Western Sahara). Algeria, Libya, Mauritania, Morocco, and Tunisia, comprise the Arab Maghreb [Western] Union
(c) '28 industrialized' states (provided here for purposes of comparison) is the mean of 28 'developed countries' for 1990, expressed in 1993 US dollars (Sivard 1993, Table II, with 1987 US$ × 1.272 = 1993 US$)

must be made adequate for all. And non-violent means of conflict resolution must be adopted, in both internal and external relations. Amicable relations between neighbouring countries have to be established and maintained for at least three essential reasons: (a) so that the land can be safely utilized; (b) so that national military expenditures can be redirected at least in part to socially and environmentally useful purposes; and (c) so that natural-resource and other environmental problems and opportunities of a multinational ecogeographical (sub-regional) nature can be addressed jointly.

9.5 Conclusion

To sum up, all foreign aid to the non-industrialized countries that attempts to ameliorate the problem of desertification—with its associated migrant problems—must without fail adopt integrated approaches which contain substantial components that: (a) address population issues; (b) support environmental education; (c) provide for the protection of biodiversity; (d) encourage participatory forms of local and national government; (e) provide opportunities for income generation outside the livestock sector; and (f) foster political security and facilitate eco-geographical (sub-regional) cooperation.

The United Nations has devoted much effort to protecting the human rights of certain categories of displaced persons (UNHCR 1993). It has as well devoted considerable effort to combating desertification in Africa and elsewhere (FAO 1993; Tolba 1991; UNSO 1992); and it is now attempting to mobilize the affected nations, in concert with the entire international community, to deal effectively with the problem (INCD 1993; UN 1993: 98–108; UNEP 1993; UNFPA 1993; UNGA 1992). It therefore becomes crucial that the United Nations deal effectively, not merely with the immediate problems, but especially with the underlying causes.

References

ACDA 1991. *World Military Expenditures and Arms Transfers 1990*. 21st edn. Washington: US Arms Control & Disarmament Agency, 148 pp.

Amer, R. *et al.* 1993. Major armed conflicts. *SIPRI Yearbook* (Oxford) 1993:81–130

Falkenmark, M. 1989. Massive water scarcity now threatening Africa: why isn't it being addressed? *Ambio* (Stockholm) 18:112–118.

FAO. 1993. *Key Aspects of Strategies for the Sustainable Development of Drylands*. Rome: Food & Agriculture Organization of the UN, 60 pp.

Fox, R. 1993. *Inner Sea: the Mediterranean and its People*. rev. edn. New York: Alfred A. Knopf, 577 pp.

Gastil, R.D. 1982. Comparative survey of freedom: the tenth year. *Freedom at Issue* [now *Freedom Review*] (New York) 1982(64):3–14.

Golini, A. et al. 1991. South-north migration with special reference to Europe. *International Migration* (Geneva) 29:253–279.

Goose, S.D. 1987. Armed conflicts in 1986, and the Iraq–Iran War. *SIPRI Yearbook* (Oxford) 1987:297–320.

Hamilton, K.A. & Holder, K. 1991. International migration and foreign policy: a survey of the literature. *Washington Quarterly* (Cambridge, MA) 14(2):195–211.

IBRD. 1992. *World Development Report 1992: Development and the Environment*. 15th edn. Washington: International Bank for Reconstruction & Development (World Bank), 308 pp.

IBRD. 1993. *World Development Report 1993: Investing in Health*. 16th edn. Washington: International Bank for Reconstruction & Development (World Bank), 329 pp.

INCD. 1993. *Elaboration of an International Convention to Combat Desertification in Countries Experiencing Serious Drought and/or Desertification, Particularly in Africa: Negotiating Text of the Convention*. New York: UN General Assembly, Document No. A/AC. 241/15, 19 November 1993, 27 pp.

Jacobsen, K. 1994. *Impact of Refugees on the Environment: a Review of the Evidence*. Washington: Refugee Policy Group, 49 pp.

Lindgren, K. et al. 1989. Major armed conflicts in 1988. *SIPRI Yearbook* (Oxford) 1989:339–355.

Loescher, G. 1992. *Refugee Movements and International Security*. London: International Institute for Strategic Studies, Adelphi Paper No. 268, 83 pp.

McColm, R.B. (ed.). 1991. *Freedom in the World: Political Rights and Civil Liberties 1990–1991*. New York: Freedom House, 479 pp.

McColm, R.B. 1992a. Comparative survey of freedom 1991–1992: between two worlds. In: McColm, R.B. (ed.). *Freedom in the World: Political Rights & Civil Liberties 1991–1992*. New York: Freedom House, 592 pp: pp 47–52.

McColm, R.B. (ed.). 1992b. *Freedom in the World: Political Rights & Civil Liberties 1991–1992*. New York: Freedom House, 592 pp.

McColm, R.B. (ed.). 1993. *Freedom in the World: the Annual Survey of Political Rights & Civil Liberties 1992–1993*. New York: Freedom House, 637 pp.

Molvær, R.K. 1991. Environmentally induced conflicts?: a discussion based on studies from the Horn of Africa. *Bulletin of Peace Proposals* [now *Security Dialogue*] (Oslo) 22:175–188.

Myers, N. 1987. Population, environment, and conflict. *Environmental Conservation* (Cambridge, UK) 14(1):15–22.

Myers, N. 1993. How many migrants for Europe? *People & the Planet* (London) 2(3):28.

Ruthström, C. 1990–1991. Refugees and European security. *Swedish Institute of International Affairs Yearbook* (Stockholm) 1990–1991:159–170.

Salih, M.M.A. 1993. Environmental conflicts in African lands: cases from Sudan and Nigeria. In: Käkönen, J. (ed.). *Perspectives on Environmental Conflict and International Relations*. London: Pinter Publishers, 162 pp: pp 116–135.

Salt, J. 1993. *Migration and Population Change in Europe*. Geneva: UN Institute for Disarmament Research, Research Paper No. 19, 81 pp.

Sivard, R.L. 1993. *World Military and Social Expenditures 1993*. 15th edn. Washington: World Priorities, 56 pp.

Suhrke, A. 1993. Pressure points: environmental degradation, migration and conflict. In: Cambridge, MA: American Academy of Arts & Sciences, International Security Studies Program, Occasional Paper Series No. 3, 67 pp: pp 3–43.

Suliman, M. 1992. *Civil War in Sudan: the Impact of Ecological Degradation*. Zürich: Swiss Federal Institute of Technology, Center for Security Studies & Conflict Research -&- Swiss Peace Foundation, Environment & Conflicts Project, Occasional Paper No. 4, 34 pp.

Tolba, M.K. 1991. *Status of Desertification and Implementation of the United Nations Plan of Action to Combat Desertification*. Nairobi: UN Environment Programme, Document No. UNEP/GCSS.III/3, 15 October 1991, xv + 88 pp + Add.1 (30 December 1991), 6 pp + Corr.1/Rev.1, 27 January 1992, 1 p.

UN. 1993. *Agenda 21: Programme of Action for Sustainable Development: Rio Declaration on Environment and Development: Statement of Forest Principles*. New York: United Nations, Publication No. DPI/1344, April 1993, 294 pp.

UNEP. 1993. *Desertification: A. Implementation of the Plan of Action to Combat Desertification in 1991–1992; B. Financing and other Measures in Support of the Plan of Action to Combat Desertification; C. Implementation in the Sudano–Sahelian Region of the Plan of Action to Combat Desertification*. Nairobi: UN Environment Programme, Governing Council Decision No. 17/19, 21 May 1993, 4 pp.

UNEP & ISRIC. 1990. *World Map of Status of Human-induced Soil Degradation*. Nairobi: UN Environment Programme, variable scale (1:4,650,000 to 1:15,000,000), 3 sheets, each 93 x 136 cm + explanatory note, 25 pp.

UNFPA. 1993. *State of World Population 1993: the Individual and the World: Population, Migration and Development in the 1990s*. New York: UN Population Fund, 54 pp.

UNGA. 1992. *Establishment of an Intergovernmental Negotiating Committee for the Elaboration of an International Convention to Combat Desertification in those Countries Experiencing Serious Drought and/or Desertification, Particularly in Africa*. New York: UN General Assembly, Resolution No. 47/188, 22 December 1992, 3 pp.

UNHCR. 1993. *State of the World's Refugees 1993: the Challenge of Protection*. Geneva: UN High Commissioner for Refugees, 191 pp. [Also: New York: Penguin Books.]

UNSO. 1992. *Assessment of Desertification and Drought in the Sudano–Sahelian Region 1985–1991*. New York: UN Sudano–Sahelian Office, 102 pp.

Wallensteen, P. & Axell, K. 1993. Armed conflict at the end of the cold war, 1989–92. *Journal of Peace Research* (Oslo) 30:331–346.

Westing, A.H. 1982. War as a human endeavour: the high-fatality wars of the twentieth century. *Journal of Peace Research* (Oslo) 19:261–270.

Westing, A.H. 1989. Environmental component of comprehensive security. *Bulletin of Peace Proposals* [now *Security Dialogue*] (Oslo) 20:129–134.

Westing, A.H. 1990. Our place in nature: reflections on the global carrying-capacity for humans. In: Polunin, N. & Burnett, J.H. (eds). *Maintenance of the Biosphere*. Edinburgh: Edinburgh University Press, 228 pp: pp 109–120.

Westing, A.H. 1991. Environmental security and its relation to Ethiopia and Sudan. *Ambio* (Stockholm) 20:168–171.

Westing, A.H. 1992. Environmental refugees: a growing category of displaced persons. *Environmental Conservation* (Cambridge, UK) 19(3):201–207.

Widgren, J. 1989. Europe and international migration in the future: the necessity for merging migration, refugee, and development policies. In: Loescher, G. & Monahan, L. (eds.). *Refugees and International Relations*. Oxford: Oxford University Press, 430 pp: pp 49–61.

Widgren, J. 1990. International migration and regional stability. *International Affairs* (Cambridge, UK) 66:749–766.

Wilson, G.K. & Wallensteen, P. 1988. Major armed conflicts in 1987. *SIPRI Yearbook* (Oxford) 1988:285–298.

WRI. 1992. *World Resources 1992–93: a Report by the World Resources Institute*. 5th edn. New York: Oxford University Press, 385 pp.

Chapter 10
Population: Perhaps the Basic Issue

10.1 Introduction

The number of humans on earth has been increasing dramatically for the past century or so, exhibiting an exponential trend that still continues.[1] Although it is not possible to establish the precise date when this number finally exceeded the global carrying capacity for humans, that time was at least several decades ago. The population transgression was thus roughly contemporaneous with the advent of the nuclear era, and vies with the latter as the most momentous event in all human history.

In the present paper I pose four questions: (1) How can it be demonstrated that the global carrying capacity for humans has been exceeded? (2) Can the global carrying capacity for humans be expanded? (3) What is the global carrying capacity for humans? (4) How is the global carrying capacity for humans to be attained?

This Chapter is reproduced from the author's Entry #203 provided in Chap. 2, with the original title, 'Our place in nature: reflections on the global carrying-capacity for humans'. It is used here by permission of the Foundation for Environmental Conservation, the copyright holder, as given on 13 March 2013 (and verified on 22 March 2013). The author is grateful to Reid A. Bryson, Carol E. Westing, and George M. Woodwell for their helpful suggestions. The **Appendix** to this Chapter is reproduced from the author's Entry #440 provided in Chap. 2, with the original title, 'All the many humans ever: an update'. It is used here by permission of the American Institute of Biological Sciences, the copyright holder, as given on 12 March 2013. The author thanks H. Erik G. Svensson for checking the data in the Appendix.

[1] The world population as at March 2013 is ca 7 billion, currently increasing at the compound rate of ca 1.1 % (giving a doubling time of ca 63 years). Cf. also Chap. 10, Appendix 10.1.

A. H. Westing, *From Environmental to Comprehensive Security*, SpringerBriefs on Pioneers in Science and Practice 13,
DOI: 10.1007/978-3-319-00687-1_10, © The Author(s) 2013

10.2 How can it be Demonstrated that the Global Carrying Capacity for Humans has been Exceeded?

Both anthropocentric and ecocentric evidence can be presented in response to this, my first question, namely, to demonstrate that the global carrying capacity for humans has been exceeded. I begin with glaring evidence in the most narrowly anthropocentric terms: today one human being out of every six is chronically undernourished, that is to say, approximately 800 million people now go to bed hungry each night (Brown 1987: 134–135; Reutlinger 1985). This shameful number of undernourished fellow humans is not only the largest such number in human history, but is, in fact, growing by about 15,000 per day.

To continue, more people by far than ever before in human history are inadequately housed; and more by far are without access to clean water. It may also be of interest to my literate audience that today there are more people by far on earth than ever before, in the full sweep of human history, who cannot read or write (a number that has been expanding steadily for about the past century). One need hardly add that this litany of depressing statistics must be viewed in the light of the truly momentous advances in science and technology that have occurred since the industrial revolution.

Further strong evidence for human overpopulation, still with a major anthropocentric component, is that, during the past decade or so, we have begun to live not only off the 'interest' of our natural heritage, but also in part off the 'principal'. Thus, the three major (and utterly essential) renewable resources—forests, grasslands, and fisheries—are, on a worldwide basis, now being utilized faster by the human species than their rate of natural replenishment (Brown et al. 1987; Holdgate et al. 1982; Mathews et al. 1986).

In yet another self-defeating action, our use of the atmosphere for the disposal of ever-increasing amounts of municipal and industrial waste products (air pollutants) is now also occurring at a rate which is faster than decomposition and dissipation can accommodate such discharges. The increasingly widespread damage to freshwater and forest ecosystems from air pollutants ('acid rain', etc.), and the dangerous build-up in the atmosphere of carbon dioxide (and other 'greenhouse' gases), provide mute testimony to this form of abuse.

A related problem awaits us as an ever-higher proportion of the world's remaining more-or-less natural areas are being converted by humans to a non-natural condition. These losses to nature will be accompanied by a loss of more-or-less subtle ecosystem services, among them: purification, hydrological, gene-bank, and aesthetic services (Bormann 1976; Ehrlich 1982; Ehrlich and Mooney 1983; Farnworth et al. 1981; Pearsall 1984; Pimental 1980; Westman 1977).

To me the most powerful evidence for human overpopulation is an ecocentric one. It concerns the adversarial relationship between our species and all the other species with which we share this earth. Our relentlessly expanding assault upon nature has both a quantitative and a qualitative component. To begin with the quantitative component: in 1850 we humans *plus* our entourage of livestock accounted for about 5 % of all terrestrial (non-marine) animal life on earth, when measured in terms of biomass, in an overall terrestrial habitat that was otherwise fully occupied by living things (Westing 1981a, b: 180). By 1950 (a mere century later), humans *plus* livestock had grown sufficiently in numbers to take over about 10 % of the world's terrestrial wildlife niche (thereby killing off the animals which we had displaced). This process of replacement—an inexorably fatal process for our wild competitors—has continued, and thus today we and our commensal animals have substituted for more than 20 % of all terrestrial wildlife. Even worse, projections indicate that in less than four decades from now this proportion of arrogation will—barring some drastic action or event—have doubled once again, to an awesome 40 %.

As for the plant life of the world, humans have to date commandeered for agriculture about 11 % of the global land area that is not continuously covered by ice (Westing 1981a, b: 180), and perhaps another 2 % for housing, transportation, and the like. In addition to using, for ourselves or our domestic animals, much of the net primary photosynthetic production occurring on the lands which we have wrested from nature for agriculture, we harvest perhaps as much as one-third of the photosynthetic production from the forests of the world, and perhaps as much as one-fourth from the grasslands (rangelands) of the world (Vitousek et al. 1986). Consider, as well, the additional photosynthetic losses to nature brought about by anthropogenic degradation of the environment (via soil erosion, desertification, air and water pollution, etc.) and by co-option of land (for cities, highways, etc.), as well as human appropriation, or prevention, of terrestrial net primary photosynthetic production: this comes today to an estimated 40 % (Vitousek et al. 1986). Overall human demands upon the green plants, the first link in the ecological chain, are thus proportionately perhaps double those of the totality of wildlife.

This massive human encroachment on terrestrial nature, which I have just outlined in rough quantitative terms for both animals and plants, also has its painful qualitative component. Thus, mammal and bird species have been driven to extinction at rates of between one and two orders of magnitude (10–100 times) greater than the normal ones (Westing 1981a, b: 180–181), and the species of other wildlife taxa will presumably be suffering extinction at comparably accelerated rates. Such losses are, of course, irreplaceable and thus unconscionable.

Before turning to the next question, it will be useful to address briefly the matter of food surpluses. It is true that a few nations produce more grain (cereals) than they can consume, and that these staple foods become export commodities. However, political and economic barriers largely prevent such food from reaching the hungry. Moreover, these food surpluses are the result of industrialized agriculture which is not merely based on special seeds, but is heavily dependent upon advanced equipment and huge inputs of fuel, fertilizer, insecticides, herbicides, and (often) artificially supplied water. In addition to being to a large extent a petrochemical (and thus non-renewable) industry, much agriculture is carried out at heavy cost to the environment.

10.3 Can the Global Carrying Capacity for Humans be Expanded?

Having established that there exist today more humans than the earth can accommodate, the second question to address is whether this dilemma could be solved by expanding the global carrying capacity for humans in lieu of contracting the population. After all (it might be said) we have been able to increase global food production in recent decades, thereby more or less closely paralleling the global population increases.

I am forced to answer this question essentially in the negative: the global carrying capacity for humans cannot be benignly expanded to any major extent. Even if the answer to this question hinged solely upon food production—which is by no means the case—the answer would still have to be essentially no, owing to the associated costs to humans and nature alluded to earlier.

There is, of course, some room for intensified agriculture leading to increased yields on a unit-area basis, but unless extreme precautions are taken, such action routinely results in damage to both humans and nature. As but one example of such harm, the so-called 'green revolution' appears to result in human deaths from pesticide poisoning in substantial excess of 10,000 per year (Loevinsohn 1987).

Some increases in food production could be achieved by not growing livestock (and pet) feed on croplands that are suitable for human food production and, to a lesser extent, by eschewing such crops as tobacco, marijuana, opium poppies, and plants destined for the manufacture of alcoholic drinks. On the other hand, the substitution of food crops for industrial crops used for fuel, fibre, or oil (cane, cotton, rape, etc.) must be denigrated because it is more sensible in many parts of the world to depend to the extent possible on renewable resources rather than on locally unavailable and expensive non-renewable oil or other mineral resources.

Only minor and localized expansion of land devoted to agriculture is still possible. Any major expansion would have to be at the expense of forestland or grassland, each of which provides services that are essential to our survival (the forestlands: wood for fuel, shelter, paper, etc.; the grasslands: livestock for meat, leather, wool, etc.; and both of them: the more subtle services alluded to earlier), and each of which is already being utilized beyond the level of sustained yield.

Moreover, expansion of croplands in most parts of the world would have to be into forestlands or grasslands that are ill-suited for agriculture, these having for the most part remained in their semi-natural state through the centuries precisely because they are agriculturally submarginal. Expansion into such lands is not only difficult and expensive, but also leads to site-deterioration (nutrient losses, soil erosion, salinization, laterization, etc.)—another self-defeating process that is becoming ever more common in many parts of the world.

A much publicized recent study by the Food and Agriculture Organization of the United Nations (FAO) should be mentioned in the present context (Harrison 1984; Higgins et al. 1982). This study suggested that global food production could be increased to serve the needs (at a minimal level of nutrition) of about twice the present global population without additional technological inputs, and of perhaps ten times the present global population with high levels of technological inputs. These conclusions are preposterous because they depend upon staple-food production on *all* land areas of the world that are theoretically capable of growing grain, whether now devoted to or occupied by grain, other food-crops, industrial crops, trees, or grass. The conclusions are further predicated upon a global distribution of food that ignores political and economic considerations. Moreover, the high-input predictions do not consider the availability, or the impacts on health and the environment, of the assigned inputs (cf. also Hekstra and Liverman 1986).

In summary, it is clear that, currently, we are not faced with a shortage of food or other resources, but rather with a 'longage' of people—to adopt a term from Garrett Hardin (Hardin 1985a: 213–215; Hardin 1986: 605).

10.4 What is the Global Carrying Capacity for Humans?

It now becomes necessary to address my third question, namely, the issue of what, in fact, the global carrying capacity for humans is: in other words, what is the maximum number of humans which the globe can accommodate?

It is quite obvious that, under certain living conditions, a human population could subsist on earth even in huge excess of today's global population. On the other hand, given that we are endowed with sapience—indeed, this property was chosen as the very epithet of our species—I am forced to equate the *maximum* number with the *optimum* number. That is to say, it is my contention that the global carrying capacity for humans must be pegged at the point at which two requisites can be simultaneously met: (1) Where everyone can have a standard of living adequate for his or her well-being (leaning here upon the Universal Declaration of Human Rights [UNGA 1948, Art. 25]); and (2) Where at the same time nature can be respected (leaning here upon the World Charter for Nature [UNGA 1982, Princ. 1]).

My approach to a first approximation of the global carrying capacity for humans has been to determine how many humans could live at the level of resource utilization or consumption that is currently enjoyed by a select group of model nations (Westing 1981). The crucial resources examined in terms of their global availability—all of which are reusable or renewable—have been five in number: (1) Total land area; (2) Cultivated land area; (3) Forestland area; (4) Grain (cereals); and (5) Wood. The nations employed in the aggregate as the model for *per caput* utilization, or consumption, have been 27 in number[2] (having an aggregate population of about 1,000 million): a diverse group accounting for all

[2] Australia, Austria, Belgium, Canada, Czechoslovakia, Denmark, Finland, France, DR [East] Germany, FR [West] Germany, Iceland, Israel, Japan, Kuwait, Libya, Luxembourg, Netherlands, New Zealand, Norway, Qatar, Saudi Arabia, Sweden, Switzerland, United Arab Emirates, United Kingdom, USA, and USSR.

those nations, the gross national product (GNP) *per caput* of which was at least twice the world average. This general approach to a determination of carrying capacity has, of course, also been used in prior instances (e.g., Hulett 1970; Odum 1970; Pirages and Ehrlich 1972; Westing 1969). The use of energy consumption as one's standard is also feasible (Hafner 1979).

It turns out that, no matter which of these five more-or-less independent resource criteria is used as the standard, the global carrying capacity for humans comes to about 2,000 million. This number might require a slight downward adjustment to the extent that the sustained global availability levels employed here have been overestimated (or might decline further in the future). On the other hand, the number might well be subject to modest and slow upward adjustments as a result of future environmentally benign or beneficial technological or political advances.

10.5 How is the Global Carrying Capacity for Humans to be Attained?

The global carrying capacity for humans today is of the order of 2,000 million. By contrast, our population today is in the neighborhood of 5,000 million—a number, moreover, that is currently increasing at the conservatively estimated compound growth rate of 1.7 %, which leads to a doubling time of a mere four decades. How, then—as my fourth and final question—is the global carrying capacity for humans to be attained?

To begin with, it must be recognized that the problem is an urgent one. The situation, already baleful, is worsening day by day and ever more rapidly. Irreversible damage is already occurring, primarily through the loss of species, and further delays in mitigating the problem will exacerbate the situation. The problem is becoming ever more intractable, and its ultimate resolution will be ever more unsatisfactory. It must further be recognized that the problem has global dimensions, many aspects of which must be dealt with at an international (or actually worldwide) level.

A three-pronged approach is required to initiate our attack on the problem: one that deals with population, another that deals with pollution, and a third that deals with habitat protection. Adjustments in population will most often directly affect developing nations; adjustments in pollution, developed nations; and adjustments in habitat protection, both developing and developed nations. However, at least the indirect ramifications of adjustments in any of these three areas will have to be internationalized to the extent possible.

Regarding *population*, growth must be arrested in whatever nation it occurs as an immediate first step towards achieving the global carrying capacity for humans. Each nation, without exception, must forthwith institute an effective programme that will maintain or reduce its birth rate to a level that, in any event, does not exceed its death rate. One sensible system that any nation could elaborate upon might involve the introduction of child-right chits (Boulding 1964: 135–136; Heer 1975; Russett 1970). In this initial step, every person who reaches his or her majority would receive from the government no more than one child-right chit, entitling the holder to have a child (with multiple births treated as a single child, and with the chit being replaced if the child fails to reach the age of one year). Such chits would be transferable (negotiable) commodities within a nation.

The right just suggested of no more than one child per person, on the average, would arrest continued population growth and would even lead to a modestly negative growth rate. However, more stringent measures would be necessary in many nations. Thus, the denomination of the child-right chit in grossly overpopulated nations might well be set at 0.5 (namely, one child permitted, on the average, for every two adults). The indicators to consider might well be staple-food production *per caput*, GNP *per caput*, rate of infant mortality, life expectancy, and literacy rate. The values of these indicators in relation to those of the aggregate of the 27 model nations leaned upon earlier, would provide one measure of overpopulation. However, worsening trends in these values within a nation would provide the most compelling demonstration of its overpopulation—as has been pointed out a number of times (e.g., Ehrlich et al. 1977: 716; Hardin 1985b: 472; Singer 1973).

Finally, in conjunction with this programme of worldwide population attrition, there would have to be established a worldwide old-age pension plan, to be financed by all nations through a progressive scheme of taxation based on the wealth of a nation. Moreover, more nearly to equalize other immediate burdens that population controls might impose upon a developing nation, a more general international fund should be established, again based upon progressive taxation, to provide further measures of relief. These could be paid out in the form of loans which would subsequently be forgiven to the extent that the necessary population goals were being achieved.

Regarding *pollution*, its global increase must also be arrested, as the second immediate step towards achieving the global carrying capacity for humans. To such ends, industrial growth must be regulated, especially in the developed nations, in some appropriate fashion so as to prevent any future increases in air pollution (cf., e.g., Broecker 1987). This action would be carried out initially in terms of national emission levels of carbon dioxide and sulphur, and perhaps also of oxides of nitrogen or other contaminants. Such a step would have to be a prelude to international cooperation in achieving, over a limited period of years, a worldwide reduction in air pollution to global levels that can be safely absorbed by the environment. Preventing further increases in the concentration of atmospheric carbon dioxide will be difficult, owing to their association with the consumption of fossil fuels, but is nevertheless feasible (Lovins et al. 1981).

To achieve safe levels of pollution discharge into the atmosphere might well require as much as, say, a reduction by one-half in the global emission rate of today—that is, perhaps a reduction to the levels of about 1965. Permitted emissions, whatever they might turn out to be, would have to be distributed to each nation on a unit-area basis. Each nation would be awarded the number of one-square-kilometer chits that corresponds to the size of its national territory (not including claimed ocean areas), such chits being negotiable commodities among nations, to the extent that they could be leased.

The necessary reductions in emissions would be achieved by a nation through the closure of industries, through restrictions on transportation, and so forth, to the extent that improvements in emission control or other means were not able to cope with the problem.

Regarding *habitat protection*, worldwide action must be taken to expand the extent of protected areas (nature reserves) as the third immediate step towards achieving the global carrying capacity for humans. Today only about 3 % of the land area of the world is reported to have a protected status (Mathews et al. 1986, Table 7.1). In only about a dozen nations is at least 10 % of the land area claimed to be protected today; and in only about 40 nations, 5 % or more. Indeed, many nations have no protected area at all. Thus, as an immediate step, each nation should be required to set aside nature reserves totalling at least 5 % of its land area (covering a range of habitat types), with the longer-term goal of at least 10 % (cf. also Harrison et al. 1984; McNeely and Miller 1983; Miller 1984; Myers 1979: 219–231). Here again, a worldwide fund should be established to facilitate this programme, with all countries contributing on the basis of a progressive scheme of taxation. Indeed, a small initial move in this direction has recently occurred (Page 1987; Walsh 1987).[3]

10.6 Conclusion

In conclusion, we have let our numbers increase recklessly, and continue to do so, with the dismaying result that we have now substantially overshot the global carrying capacity for humans. This increase in our numbers and accompanying demands occurs at the direct expense of the other living things on earth: our *over* population is inexorably coupled with their *under* population and, sad to say, their disappearance. Our behavior as a species is both wanton and stupid: *wanton* because we are callously vandalizing nature; and *stupid* because we cannot lead a healthy and satisfying life in the absence of a clean, productive, and aesthetically pleasing, environment. Indeed, we will not even be able to survive in an environment that is as polluted, impoverished, and unstable as ours is on its way to becoming. Moreover, it is not possible for us to plead ignorance of our unethical and self-defeating approach towards nature, given such erudite and eloquent recent expositors—or, must I say, Cassandras?—as Lester Brown (Brown 1978), David

[3] Formally protected areas worldwide as of September 2012: 1.6 % of the ocean (although ca 7 % of territorial waters), and 12.7 % of the land (cf. http://www.unep-wcmc.org/ppr2012_903.html).

Ehrenfeld (Ehrenfeld 1978), Garrett Hardin (Hardin 1985a), and Paul Ehrlich (Ehrlich 1986).[4]

In order for all of us—that is, for all humans on earth—to be able to live within the constraints of the global environment, at a level of health and well-being which is equivalent to that of the 2,000 million of the most affluent of us, we need to reduce our numbers substantially and without delay. Alternatively, we might well all live a life that is less affluent (some would say, less profligate) than this, necessitating little more than a modest reduction in our present numbers.

The application of technological advances in resource utilization (including energy production), pollution abatement, and site rehabilitation, will lead to more or less modest increases in the global carrying capacity for humans. A crucial component of such advances will be the adoption of more conservative life-styles (energy conservation, material—including water—conservation and recycling, etc.). Conversely, future shortages in non-renewable resources will lead to more or less modest decreases (Cook 1976). The net effect on the global carrying capacity for humans in the years to come is difficult to predict, but may well be positive.

The political and other social problems that must be overcome to attain the necessary goals have not been addressed in my present reflections, but will be formidable indeed (cf., e.g., Catton 1980; Slesser 1972), not least among them being the decade or more which it can take for new ideas to be accepted (Barber 1961; Deutsch et al. 1971), and the need for global cooperation. Among other things, we must also finally learn not to damage the environment and squander its resources on such frivolous human pastimes as war and the preparation for war (Westing 1980; Westing 1983; Westing 1986). All of our actions must be judged not only by literate standards, but also by 'numerate' and 'ecolate' ones (Hardin 1985a; Opschoor 1987). We must educate ourselves to accept new ways of thinking and adopt new cultural norms that: (1) recognize the necessity of managing our renewable natural resources on a basis of sustained yield; (2) recognize the importance of thinking in terms of future generations; and (3) recognize the rights of the other living things with which we share this earth.

In short, we must develop a true sense of our place in nature.

Appendix 10.1 All the Many Humans Ever[5]

Some 30 years ago a poem, in voicing its author's distress with our overcrowded planet (4.4 billion people alive at the time), bolstered his concern by claiming that '...there are now more of us alive than ever have been dead. I don't know what this

[4] The most recent warning by Lester R. Brown is in his *World on the Edge: How to Prevent Environmental and Economic Collapse* (New York: W.W. Norton, 240 pp, 2011). And the most recent one by Paul R. Ehrlich is in *Science* (Washington) 340(6130):324–328; 19 April 2013.

[5] Cf. this Chapter's unnumbered opening note.

means, but it can't be good' (Matthews 1979: 36). A colleague asked me to verify that disquieting assertion.

My subsequent stab at estimating the numbers of humans who had ever lived (i.e., for the then past 300 thousand years) came to approximately 46.4 billion. That is to say, those living at the time were thus not even close to the assertion in question—rather representing only 9 % of all the humans ever born (Westing 1981–1982).

My analysis necessitated a number of assumptions:

1. ***As to when it all began***, I decided that *Homo sapiens* evolved from *H. erectus* and *H. habilis* in about 298,000 BCE, basing this on a separate detailed analysis of the matter (Campbell 1974: 111).
2. ***As to the seven ages (and eight dates) I established for my calculations***, their transitions were based on what seemed to me to be key events in human history, moreover, for which reasonably reliable population numbers had been established (Coale 1974). [It could be noted here that the equations presented in my earlier analysis (Westing 1981–1982, Table 2) can be used for obtaining a rough population estimate for any date in human history.]
3. ***As to the increasing life spans I employed for my seven ages***, these were simply educated guesses on my part, based on what I could glean from the relevant literature, and not including neonatal deaths.

As an aside, it may be of interest to note that somewhat earlier, it had been suggested that, 'Behind every man now alive stand thirty ghosts, for that is the ratio by which the dead outnumber the living. Since the dawn of time, roughly a hundred billion human beings have walked the planet Earth' (Clarke 1968: Foreword, p. 1). Unlike the poem that started all this, that guess at least erred in the right direction. However, at the time of that statement, 1968, there lived about 3.5 billion people from among an all-time total at the time (employing my assumptions) of about 45.6 billion, thus coming to only 13 ghosts behind each person then alive (i.e., about 8 % of them). On the other hand, I might mention that a recent attempt at estimating these values (based on assumptions and a mathematical approach quite different from mine) comes quite close to that conjecture (Haub 2002).

Updating the above 1980 values to 2010 (and employing the same methodology as before), the number of people worldwide to have ever been born rises to approximately 49.2 billion. With the United Nations global population estimate for mid 2010 being about 6.9 billion, those of us living today thus now represent the somewhat higher fraction of about 14 % of all those who have ever lived.

Sad to say, we continue to ignore the ever growing multiplicity of indications that the earth's capacity to sustain a healthy biosphere is being increasingly overwhelmed by our relentlessly expanding human numbers, needs, and desires (e.g., Westing 2010).

References

Barber, B. 1961. Resistance by scientists to scientific discovery. *Science* (Washington) 134:596–602.

Bormann, F.H. 1976. Inseparable linkage: conservation of natural ecosystems and the conservation of fossil energy. *BioScience* (Reston, VA) 26:754–60.

Boulding, K.E. 1964. *Meaning of the Twentieth Century: The Great Transition*. New York: Harper & Row, 199 pp.

Broecker, W.S. 1987. Unpleasant surprises in the greenhouse? *Nature* (London) 328:123–126.

Brown, L.R. 1978. *Twenty-ninth Day: Accommodating Human Needs and Numbers to the Earth's Resources*. New York: W.W. Norton, 63 pp. 5.

Brown, L.R. 1987. Sustaining world agriculture. In: Brown, L.R. et al. 1987. *State of the World 1987: a Worldwatch Institute Report on Progress Toward a Sustainable Society*. New York: W.W. Norton, 268 pp: pp 122–138, 235–238.

Brown, L.R. et al. 1987. *State of the World 1987: a Worldwatch Institute Report on Progress Toward a Sustainable Society*. New York: W.W. Norton, 268 pp.

Campbell, B.G. 1974. *Human Evolution: an Introduction to Man's Adaptations*. 2nd edn. Chicago: Aldine Publishing Co., 469 pp. [Now in its 1998, 4th edn.]

Catton, W.R., Jr. 1980. *Overshoot: the Ecological Basis of Revolutionary Change*. Urbana, IL: University of Illinois Press, 298 pp.

Clarke, A.C. 1968. 2001: *a Space Odyssey*. New York: New American Library, 221 pp.

Coale, A.J. 1974. History of the human population. *Scientific American* (New York) 231(3):40–51,210.

Cook, E. 1976. Limits to exploitation of nonrenewable resources. *Science* (Washington) 191:677–682.

Deutsch, K.W. et al. 1971. Conditions favoring major advances in social science. *Science* (Washington) 171:450–459.

Ehrenfeld, D. 1978. *Arrogance of Humanism*. New York: Oxford University Press, 286 pp.

Ehrlich, P.R. 1982. Human carrying capacity, extinctions, and nature reserves. *BioScience* (Reston, VA) 32:331–333.

Ehrlich, P.R. 1986. *Machinery of Nature*. New York: Simon & Schuster, 320 pp.

Ehrlich, P.R. & Mooney, H.A. 1983. Extinction, substitution, and ecosystem services. *BioScience* (Reston, VA) 33:248–254.

Ehrlich, P.R. et al. 1977. *Ecoscience: Population, Resources, Environment*. San Francisco, CA: W.H. Freeman, 1051 pp.

Farnworth, E.G., et al. 1981. Value of natural ecosystems: an economic and ecological framework. *Environmental Conservation* (Cambridge, UK) 8(4):275–282.

Hafner, E. 1979. Human equivalent power: towards an optimum energy level, *Environment* (Philadelphia, PA) 21(6):34–37.

Hardin, G. 1985a. *Filters Against Folly: How to Survive despite Economists, Ecologists, and the merely Eloquent*. New York: Viking Penguin, 240 pp.

Hardin, G. 1985b. Human ecology: the subversive, conservative science. *American Zoologist* (USA) 25:469–476.

Hardin, G. 1986. Cultural carrying capacity: a biological approach to human problems. *BioScience* (Reston, VA) 36:599–606.

Harrison, J. et al. 1984. World coverage of protected areas: development goals and environmental needs. In: McNeely, J.A. & Miller, K.R. (eds). *National Parks, Conservation, and Development: the Role of Protected Areas in Sustaining Society*. Washington: Smithsonian Institution Press, 825 pp: pp 24–33.

Harrison, P. 1984. *Land, Food and People*. Rome: Food & Agriculture Organization of the UN, 96 pp+4 maps.

Haub, C. 2002. How many people have ever lived on earth? *Population Today* (Washington) 30(8):3–4.

Heer, D.M. 1975. Marketing licenses for babies: Boulding's proposal revisited. *Social Biology* (Madison, WI) 22:1–16.
Hekstra, G.P. & Liverman, D.M. 1986. Global food futures and desertification. *Climatic Change* (Dordrecht) 9:59–66.
Higgins, G.M. et al. 1982. *Potential Population Supporting Capacities of Lands of in the Developing World.* Rome: FAO, 139+6 pp.
Holdgate, M.W. et al. (eds). 1982. *World Environment 1972–1982: a Report by the United Nations Environment Programme.* Dublin: Tycooly International, 637 pp.
Hulett, H.R. 1970. Optimum world population. *BioScience* (Reston, VA) 20:160–161.
Loevinsohn, M.E. 1987: Insecticide use and increased mortality in rural central Luzon, Philippines. *Lancet* (London), 1987(2):1359–1362.
Lovins, A.B. et al. 1981. Least-cost Energy: Solving the Co_2 Problem. Andover, MA: Brick House, 184 pp.
McNeely, J.A. & Miller, R.K. 1983. IUCN, national parks and protected areas: priorities for action. *Environmental Conservation* (Cambridge, UK) 10:13–21.
Mathews, J.T. et al. 1986. *World Resources 1986: a Report by the World Resources Institute and the International Institute for Environment and Development.* New York: Basic Books, 353 pp.
Matthews, W. 1979. Inheritance tax. *Vegetable Box* (Seattle) 1979(4):36–37.
Miller, K.R. 1984. Bali action plan: a framework for the future of protected areas. In: McNeely, J.A. & Miller, K.R. (eds). *National Parks, Conservation, and Development: the Role of Protected Areas in Sustaining Society.* Washington: Smithsonian Institution Press, 825 pp: pp 756–764.
Myers, N. 1979. *The Sinking Ark: a New Look at the Problem of Disappearing Species.* Oxford: Pergamon Press, 307 pp.
Odum, E.P. 1970. Optimum population and environment: a Georgian microcosm. *Current History* (Philadelphia, PA) 58:355–359,365–366.
Opschoor, J.B. 1987. Ecologising socio-economic and scientific progress. *Scientific World* (London) 31(3):18–20.
Page, D. 1987. Cutting the debt, saving the forest. *Environment* (Philadelphia, PA) 29(7):4–5.
Pearsall, S.H., III. 1984. *In absentia* benefits of nature preserves: a review. *Environmental Conservation* (Cambridge, UK) 11:3–10.
Pimental, D. 1980. Environmental quality and natural biota. *BioScience* (Reston, VA) 30:750–755.
Pirages, D. & Ehrlich, P.R. 1972. If all Chinese had wheels. *New York Times* 121(41,690):47. 16 March 1972.
Reutlinger, S. 1985. Food security and poverty in LDCs. *Finance & Development* (Washington) 22(4):7–11.
Russett, B.M. 1970. Licensing: for cars and babies. *Bulletin of the Atomic Scientists* (Chicago) 26(9):15–9.
Singer, S.F. 1973. Possibilities and usefulness of defining optimum population levels. *International Journal of Health Services* (Farmingdale, NY) 3:617–622.
Slesser, M. 1972. *Politics of the Environment: Including a Guide to Scottish Thought and Action.* London: George Allen & Unwin, 176 pp.
UNGA. 1948. *Universal Declaration of Human Rights.* New York: UN General Assembly Resolution No. 217(III)A, 10 December 1948, 3 pp.
UNGA. 1982. *World Charter for Nature.* New York: UN General Assembly Resolution No. 47/7, 28 October 1982, 5 pp.
Vitousek, P.M. et al. 1986. Human appropriation of the products of photosynthesis. *BioScience* (Reston, VA) 36:368–373.
Walsh, J. 1987. Bolivia swaps debt for conservation. *Science* (Washington) 237:596–597.
Westing, A.H. 1969. How dense can Vermonters get? *Vermont Freeman* (Plainfield, VT) 1(22):10–11.
Westing, A.H. 1980. *Warfare in a Fragile World: Military Impact on the Human Environment London*: Taylor & Francis, 249 pp.

Westing, A.H. 1981. A world in balance. *Environmental Conservation* (Cambridge, UK) 8(3):177–183.

Westing, A.H. 1981–1982. A note on how many humans that have ever lived. *BioScience* (Reston, VA) 31(7):523–524; 32(1):6.

Westing, A.H. 1983. Pursuit of security: how long can the planet stand it? *Development Forum* (United Nations, Geneva) 11(3):1,10. April 1983.

Westing, A.H. 1986. Expanded concept of international security. In: Westing, A.H. (ed.). *Global Resources and International Conflict: Environmental Factors in Strategic Policy and Action.* Oxford: Oxford University Press, 280 pp: pp 183–200.

Westing, A.H. 2010. Overpopulation and climate change. *International Herald Tribune* (Paris) 2010(39,482):6. 18 February 2010.

Westman, W.E. 1977. How much are nature's services worth? *Science* (Washington) 197:960–964.

Units of Measure

Standard International (SI) prefixes used in the text:

T (Tera) = 10^{12}	c (centi) = 10^{-2}
G (Giga) = 10^{9}	m (milli) = 10^{-3}
M (Mega) = 10^{6}	μ (micro) = 10^{-6}
k (kilo) = 10^{3}	n (nano) = 10^{-9}

Degrees Celsius (°C) = a measure of temperature. To convert degrees Celsius to degrees Fahrenheit (°F), first multiply by 1.8, then add 32

hectare (ha) = 10^4 m^2 = 0.01 km^2 = 2.471 acres

hertz (Hz) = a measure of frequency in numbers per second (s^{-1})

kilogram (kg) = 2.205 pounds

kilogram per hectare (kg/ha) = 0.892 pound per acre

kilometer (km) = 10^3 m = 0.621 statute mile = 0.540 nautical mile

kilometer, square (km^2) = 10^6 m^2 = 100 hectares (ha) = 247.1 acres = 0.386 square statute mile

meter (m) = 3.281 feet

meter, cubic (m^3) = 10^3 liters (L) = 264.2 US gallons = 220.0 British gallons = 6.290 US oil barrels = 0.000811 acre-foot

A. H. Westing, *From Environmental to Comprehensive Security*,
SpringerBriefs on Pioneers in Science and Practice 13,
DOI: 10.1007/978-3-319-00687-1,

Glossary

AAAS American Association for the Advancement of Science, Washington (1848–)

ACDA US Arms Control and Disarmament Agency, Washington (1961–1997)

CMEA Council for Mutual Economic Assistance ('COMECON'), Moscow (1949–1991)

CoE Council of Europe, Strasbourg, France (1949–)

FAO Food and Agriculture Organization of the United Nations, Rome (1945–)

GEF Global Environment Facility, Washington (1990–), a joint UN agency of IBRD(WB), UNDP, and UNEP

IBRD(WB) International Bank for Reconstruction and Development, Washington (1945–); most often referred to as the World Bank

ICRC International Committee of the Red Cross, Geneva (1863–)

IUCN International Union for Conservation of Nature, Gland, Switzerland (1948–); for a time referred to as the World Conservation Union

LNTS *League of Nations Treaty Series*, New York (1920–1946) [cf. UNTS below]

NATO North Atlantic Treaty Organization, Brussels (1949–)

Number [The thus numbered references throughout the text are provided in Chapter 2.]

PRIO International Peace Research Institute Oslo, Oslo (1959–)

SIPRI Stockholm International Peace Research Institute, Stockholm (1966–)

UNCSD United Nations Commission on Sustainable Development, New York (1992–)

A. H. Westing, *From Environmental to Comprehensive Security*,
SpringerBriefs on Pioneers in Science and Practice 13,
DOI: 10.1007/978-3-319-00687-1,

UNEP United Nations Environment Programme, Nairobi (1972–)

UNESCO United Nations Educational, Scientific and Cultural Organization, Paris (1948–)

UNFPA United Nations Population Fund, New York (1967–); originally named the United Nations Fund for Population Activities

UNHCR United Nations High Commissioner for Refugees, Geneva (1950–); occasionally referred to informally as the United Nations Office for Refugees

UNIDIR United Nations Institute for Disarmament Research, Geneva (1980–)

UNWCED United Nations World Commission on Environment and Development ('Brundtland Commission'), Geneva (1983–1987)

UNTS *United Nations Treaty Series*, New York (1946–) [cf. LNTS above]

WWF World Wide Fund for Nature, Gland, Switzerland (1948–); originally named the World Wildlife Fund

About the Author

Westing's undergraduate training was in botany (Columbia, AB, 1950). After two years in the United States Marine Corps (serving as an artillery officer in the Korean War) he became a forest ecologist (Yale, MF, 1954; Ph.D., 1959). He has been a Research Forester with the United States Forest Service, has taught forestry, ecology, and conservation at various colleges and universities, was Dean of Natural Science at Hampshire College, has twice been a Research Fellow at Harvard (Bullard, Guggenheim), and has been a Senior Researcher at the Stockholm International Peace Research Institute (SIPRI) as well as the Peace Research Institute Oslo (PRIO). For eight years he directed the United Nations Environment Programme (UNEP) project on 'Peace, Security, & the Environment', a position which took him to many countries throughout the world; and is the author of numerous articles and several books in that subject area. Westing has been on the faculty of the European Peace University (EPU), a member of the International Union for Conservation of Nature (IUCN) World Commission on Protected Areas, Vice-President of the International Society of Naturalists (INSONA), and also a member of or advisor to a number of other international environmental nongovernmental organizations and scholarly journals. He has served on the Board of the Korean DMZ Forum. He was awarded an honorary doctorate (DSc, Windham, 1973) and medals from the New York Academy of Sciences (1983) and Government of Bulgaria (1984); he was named a 'Peace Messenger' (together with four international colleagues) by the United Nations Secretary-General (1987), and he is one of the 500 individuals worldwide to have been appointed to the United Nations 'Global 500 Roll of Honour' (1990). He has been a Consultant in Environmental Security since 1990, variously to the World Bank, UNEP, UNIDIR, and UNESCO, to the International Committee of the Red Cross (ICRC), to the International Organization for Migration (IMO), the Government of Eritrea, and to several other national and international agencies.

A. H. Westing, *From Environmental to Comprehensive Security*,
SpringerBriefs on Pioneers in Science and Practice 13,
DOI: 10.1007/978-3-319-00687-1, © The Author(s) 2013

About the Book

This work presents the evolution of the traditional concept of 'national security' as military security to additionally embrace 'environmental security' and then necessarily also 'social (societal) security', thence to be termed 'comprehensive human security'. It accomplishes this primarily by presenting 11 of the author's own benchmark papers published between 1986 and 2010 (additionally providing bibliographic citations to a further 35 of the author's related publications during that period). The work stresses the importance of transfrontier (regional) cooperation, and also recognizes global overpopulation as a key impediment to achieving comprehensive human security.

A. H. Westing, *From Environmental to Comprehensive Security*,
SpringerBriefs on Pioneers in Science and Practice 13,
DOI: 10.1007/978-3-319-00687-1, © The Author(s) 2013

MIX
Papier aus verantwortungsvollen Quellen
Paper from responsible sources
FSC® C105338

If you have any concerns about our products,
you can contact us on
ProductSafety@springernature.com

In case Publisher is established outside the EU,
the EU authorized representative is:
Springer Nature Customer Service Center GmbH
Europaplatz 3, 69115 Heidelberg, Germany

Printed by Libri Plureos GmbH
in Hamburg, Germany